From a small town, Callan, County Kilkenny, in the Republic of Ireland, Philip William Bryan currently lives in the city of Guiyang in the Guizhou Province of the People's Republic of China, where he works as an English and a physical education teacher. He is currently working at the Number One High School of Guiyang. He spent many years prior working in the retail industry specifically managing forecourt and service stations.

After a life-changing trip to the Hashemite Kingdom of Jordan in 2015, he has changed his lifestyle around completely. He no longer is a prisoner of the Monday to Friday nine to five routine but now enjoys a rewarding and fulfilling life in China, where he has been since 2016. His travel stories are based on his own experiences and life.

Ever since that trip to Jordan and the inspiring people he met there, his attitude and way of life have changed dramatically.

There are so many people who I would like to dedicate this book to. I would like to dedicate it to the following people.

To the memory of my dear grandmother, Josie, grandaunt, Marie, cousin, Val, and aunt, Veronica, who were always there for me. Treasured memories that are cherished. You are loved dearly and missed every day.

I dedicate this book to you—the reader. My goal for this book is that I hope it can inspire and motivate you to get out and explore our amazing world, try new things, embrace different cultures, and meet new people. Step outside and you will see just a piece of sky. There is more, so much more, do not settle for that small piece. Get out, grab life, live, experience and love.

—Philip

Philip Bryan

SQUAT TOILETS AND CHOPSTICKS

Experiencing Life Through Travel

AUSTIN MACAULEY PUBLISHERS™

LONDON • CAMBRIDGE • NEW YORK • SHARJAH

A CIP catalogue record for this title is available from the British Library.

ISBN 9781528953320 (Paperback)
ISBN 9781398401464 (ePub e-book)

www.austinmacauley.com

First Published 2024
Austin Macauley Publishers Ltd®
1 Canada Square
Canary Wharf
London
E14 5AA

There are so many people that I wish to thank who have inspired me throughout my life. If I were to list everyone, I would need a separate book for that.

There are some that I wish to acknowledge.

To my parents, Helen and Kevin. I would not be the person I am today or do the things I am doing if it were not for your love and support.

To my uncle Sean and uncle Roy. You both inspire and motivate me so much.

To Austin Macauley Publishers. I am so grateful to you for helping me achieve one of my life goals. Thank you for this amazing opportunity.

To Dorothy, who helped me during my hardest time. I will always appreciate what you did for me.

To the group of wonderful and inspirational people I travelled around Jordan with, thank you for opening my eyes and showing me that I could live an amazing life.

To the amazing students I had the opportunity to teach on Sunday afternoons, Tom, Audrey, Toby, Brian, thank you for inspiring me.

To the people who doubted me and told me that I would never accomplish anything in life, I took great pride in proving you wrong.

Finally, to the most amazing friends: Xiao RenFei (Brian), Xu Haonan, (Rick), Courtney M. Jones, Paul Chamless, Riley Mills, Lin Ziyue (Rebecca), Zhang PeiYu (Joyce), Billy Brennan, Su Zifan (Milo), Yang Jie, Li Yang (Leon), Magda Greco, Beverly Hermosilla Serrano, Sinead Brennan, Stephen Kenna, Sean Kenna, Lyza O'Halloran, Alice Lacey, Ruthnie Benoit, Fraser Mitchell, Michelle Stapleton, Padraic Bacon, and the wonderful colleagues I work with at No. 1 High School in Guiyang.

Without the help and support of you, there would be simply no way that I could live the life I am living, be the person I am, or doing the things I am doing without having you there for me. Friends are the family we get to choose for ourselves. Thank you.

Table of Contents

Introduction..11

Chapter One: A Piece of Sky................................14

Chapter Two: I Sent an Email................................19

Chapter Three: Losing My Travel Virginity..................28

Chapter Four: What Brought Me to Now......................39

Chapter Five: Chasing Rainbows.............................66

Chapter Six: A Change of Plan...............................73

Chapter Seven: Isolation 83

Chapter Eight: Out of My Control............................90

Chapter Nine: Out the Window...............................104

Chapter Ten: The Teacher Is Also the Student 110

Chapter Eleven: Lao Wai 118

Chapter Twelve: Culture Shock 128

Chapter Thirteen: Through My Eyes..........................142

Chapter Fourteen: Take the Good with the Bad 152

Chapter Fifteen: Goin' Back 167

Conclusion..176

Introduction

I could never imagine myself writing a book about my travel experiences. It was something that never crossed my mind. My initial thoughts were 'who would read this?' A good friend of mine messaged me when I was in Israel in the Summer of 2016, and she told me that I should start writing down my travel experiences because people would find them interesting.

I started thinking about doing this. It was not until the beginning of 2017, that I began writing down my experiences from different countries that I had been to up to that point, so then I began writing in a journal that a friend bought me for my birthday.

It did not take long until I started to enjoy writing about my different experiences in different places and I was filling up journals and books with different tales and stories from my travels.

I am by no means a professional writer. I am simply an English teacher in China who loves to travel and see cool things and I write only about my own personal experiences. I decided then that if I were to write a book, it would not be useful as a guide to visit places, there are too many books like that out there. But rather I would write a book which takes readers on my travels with me to different places and lets them feel and experience what I felt and saw.

I want the reader to feel emotions. Emotions that were felt when I went to Cambodia, Jordan, Israel, New York, and China. A lot of this book will focus on the impact of my trip to Jordan, and it will also focus on my life in China. What motivated me to come here and where I am still today.

This book will take us on two journeys, the first being the different destinations that I have had the pleasure of travelling to, and the second being the journey that made me, and brought me to where I am now.

Just like with any story, it is always best to start at the beginning, so this book begins with my earlier travel experiences when I was a teenager. You will learn about my struggles that brought me to where I am now and that I was in a soul-destroying job that gave me severe anxiety and stress and inevitably drove me to find a new career in teaching and living a more fulfilling life in China.

Throughout my journey in China, you the reader will learn about life here, how I became a student, and my students became my teachers. In addition to this, I did not just learn about different traditions and cultures, but I also learned so much about myself on this journey.

When I made the decision to write this book, I wrote it with an intention to inspire readers and dreamers. You, wonderful people who are looking for something more than the little piece of sky you see outside your window. I hope you can learn from my experience and live the life you always dreamed about living. I know I once thought about the life I now have and could not have imagined doing what I am doing, working in a rewarding job, and travelling to amazing places.

So, I hope this book will not only take you on my journey, but I really want it to inspire you to begin your own journey to freedom and live on your own terms. I look forward to taking you on my travels.

Exploring

Chapter One

A Piece of Sky

I guess I have always enjoyed travelling. I loved short trips that took me to places such as London, New York, and Barcelona. But then I started to enjoy longer trips to Australia, and to the Hashemite Kingdom of Jordan. Australia, I guess was the trip that tested me to see if I would enjoy staying away for longer periods than a week. Up until that point, I had not been outside of my country for more than eight days. So, in one way, going to Australia was a test to me to see if I could cope with being away for a long period of time.

I was about to embark on a trip for a duration of three weeks. This test provided me with a positive outcome, which was the fire and desire to travel more was beginning to grow inside of me. But I must confess and state that it was not always like that for me. I will share with you two stories from my youth.

These two stories will define me as a home bird…

A home bird or a home body, according to Collins Dictionary, is a person who 'is reluctant to leave their hometown or childhood home, or who returns after a period of living away'.

The first trip I ever went on was back in 1999, when my family and I went to visit my dad's younger sister Gabrielle, in London. I remember earlier on in that year; she was over with us for my brother's confirmation. After that she sat down with my parents and planned a week's vacation over with them. When my parents told us that we would be going over to London in July for a week, what did I do?

I did not get excited, nor did I jump with joy at the thought of being on my first ever plane ride. Instead, I ran upstairs, hid in a closet and cried! I was just nine years old. I was crying so much on my own there in the closet that I couldn't speak, instead I used chalk and wrote on the door that there was no way I was going to go to London or in fact ever go anywhere.

Secondly, I remember when I was 13 years old, I was in secondary school. We went away for a weekend to the Irish speaking community in the Southwest Region of Ireland with the school, and I do recall on the Sunday afternoon while in Tralee, County Kerry, I was crying in a park, simply because I was homesick for my family, my mother, and my father. Even though my brother was on the same trip with me for the few short days, I still felt uncomfortable. To also make this feeling worse for me, my mam and dad decided to go to Killarney, which was roughly one hour away from where we were in rural Ireland, so what was the problem with me? It was simple I just had never been away from home for any period! Not without my family anyway! Up until that time the most I was away without my parents for was a sleepover at a friend's house which was five minutes from my home and in the same town as where I lived!

So, without a doubt you see me as a home bird, someone who would never ever venture any further than Callan, let alone Ireland.

What happened? What opened my eyes to see a bigger world? What pushed me to want more and to see more?

What gave me that magnificent infection people call 'the travel bug'? I will share with you a very brief story of how it came to infect me so positively.

After working a terrible and soul-destroying summer job in 2012, I saved enough money to be able to do something amazing. A once in a lifetime thing. Not just going to New York, I could go to New York anytime, but going to New York was for a concert. Now, you may laugh at this and it's okay, not everyone is meant to be the same nor share the same interests. I was going to see Barbra Streisand perform for the first time in her native home of Brooklyn. Never had she performed a concert there and this, in my opinion, was going to be a historic event for one of perhaps the greatest performers that has ever graced the stage.

One thing that separated this trip away from everything else was simply that I was going on my own! My very first solo trip and it was to New York. Never had I been anywhere that meant I was on my own. For the first time, I had control over every single aspect of my trip, the hotel, the flights, the attractions and most importantly the time. I decided exactly what to do with it, I did not have to consult anyone, nor did I have to do the things I did not want to do. I was the commander, the creator of my own path and experience, and it was me who decided what step to take.

Every experience we have is a lesson. I believe that every situation, everything we face is either a teacher or a test. A teacher does not have to be a human being in a classroom or a lecture hall. Quite simply a teacher can be a person, a place or a situation that gives us the opportunity to grow or to learn in any shape or form. It presents us with both positive and negative solutions, thoughts, ideas, and experiences. The homework that this 'universal teacher' gives us is more of an independent theorising assignment. We need to decide for ourselves what we learn from it whether we take it good or bad, it is up to our own interpretation. There are no mistakes, just lessons.

Honestly, that solo trip to New York was my wake-up call that I wanted to travel more, and it was my own personal test, my challenge. And was I able to pass it? Absolutely, I passed that bitch with flying colours. I graduated with two things, my wings and a want that will never be satisfied.

My visit to the Hashemite Kingdom of Jordan was the trip that made the travel bug in me terminal. What I mean is that Jordan was the awakening point in my life where I realised that I needed more, more travel, more experience and so many more opportunities in my life. These wants, these needs could not be realised or even visualised from my desk in the gas station that I was managing at that time.

I believe I was stuck in a rut, in a dead-end job, even though I was working for the biggest private owned company in Ireland at the time. You would think I would be proud to be presented with so many opportunities in life. No. Instead I was presented with anxiety, stress, and depression. Jordan ruined me in so many wonderful ways. It was not just the landmarks, the culture or the history that changed me. But in fact, it was the people who I travelled with. It was not a solo trip but a group trip. People from all over the globe had assembled to embark on this eight-day-long excursion of the ancient land.

When I say the people, I simply mean these amazing people who inspired me so much, more than they will ever know. Some were just enjoying a break from their everyday life, and some were simply living their dreams. One of them was working in London and had finished up and was now returning home to Australia, but not before he could spend several months backpacking around so many different countries. Another was a girl from New York, who really pushed herself out of her comfort zone by embarking on her first major trip so far east, and finally a couple who took every chance they had to get out of their '9–5 prison' to get out and see the world. These four individuals are half the reason why I am doing what I am doing now in my life. They made me realise that there was so much more to life than working in a job that I had grown to detest.

Since Jordan, I have so proudly welcomed them to my home, and my country and developed lifelong friendships with them simply because we enjoyed and experienced just a few short days together in a country. This is what travelling is about, it is not always about the places, but I believe it is also about the people you meet on the way.

To the people who changed my life, I express my sincere gratitude and thanks for helping me realise that what I was looking at in the sky was simply a small piece and that there was so much more to see.

Chapter Two

I Sent an Email

In the fall of 2005, my nan bought a new computer for my brother and me. This also meant we could finally get internet in our house, although it was the old dial-up connection which meant you couldn't use the telephone while you were on the internet, but it was better than nothing. It was also at this time I set up my first email account and I started to email my aunt Molly, who lived in Windsor, England. She told me to get one and that way we could stay in touch more frequently.

The months had passed, and we were exchanging emails back and forth to one another. Then one night she sent me one of those chain emails, that you must send to everyone in your inbox. As I read the attachment, I saw a list of other people's email addresses that she had forwarded it on to, on this list I saw the names Edith and Sean.

There were eight children born to Jack and Molly Bryan three sons: Michael, Sean, Kevin, and five daughters: Margaret, Angela, Molly, Veronica, and Gabrielle (the first four were born in London, the second four born in Ireland, and there are some age gaps between the first and the other four). Sean is the third oldest in the family and he is fourteen years older than my father. When Dad was growing up, Sean was leaving for England. This age gap between them meant that neither of them got to know each other, both in their earlier and later lives. They both had their own lives and there was not much communication between them. The last time they spoke prior to 2006 was at their mother's funeral in 1992.

After I saw their email address, I made the decision to contact them. I emailed Edith and introduced myself, almost like a letter to a potential pen pal. I was excited to be contacting her and Sean. The next evening after coming home from school, I signed into my email and there it was, Edith's reply. She had sent it at 6 am that morning. I was so happy to hear from her. I had started something. I had discovered a part of my family that I knew nothing about.

"Dear Philip,

It was a nice surprise to hear from you. I am not surprised that you do not remember meeting us so many years ago, you were just a baby. As you probably know, I have not been well lately, but I am in good spirits, and I am sure the good Lord has his hands on me. Uncle Sean is doing well, he is over at our daughter Debbie's house doing some work in her garden. She has many chickens whose ancestors were hatched from eggs that came from Callan, many years ago. He goes there about three times a week. Debbie has a beautiful garden, and she grows many fruits and vegetables that she sells at the church gate every Sunday. We used to have a table outside her house, but it was being robbed so we had to remove it. I must have more treatment weeks, so I will not be able to reply to you until I feel well. But I promise always to reply to you, please give your parents our love. Hope to hear from you soon.

Love Edith and Sean"

I was overwhelmed by this response, and I replied immediately. This was April 2006, the month when I was beginning to build a new relationship with a part of my family that remained unknown. Over the course of the next few months, we would write to each other talking about our lives and what we were doing. We also exchanged stories and photographs.

On August 5, 2006, the eldest of Dad's siblings, his brother Michael, died due to complications after having an operation on his bowel. He was especially important to both my brother and me because he was the only biological uncle that we grew up knowing. He was always Uncle Michael, and we took his death extremely hard. I remember when I was emailing Edith during these days, she sent me a lovely message sympathising. But, due to her deteriorating health neither Edith nor Sean were able to come to the funeral.

Throughout the remainder of August and early September, our emails to each other became fewer and fewer due to her health. But through our messages and chain mails, I contacted her youngest daughter, Sinead through email and then I started to message her. We were so happy to contact each other. To my surprise, Sinead and her family were coming to Ireland the end of October for a few days. We were able to discuss dates and organise for her to come and see us.

Before she came to Ireland, I decided that I would write a letter to Edith because I had not heard from her for a while. Her condition was worsening.

I wrote:

"Dear Aunt Edith,

I am sorry to hear you are not well. I wanted to write to you to tell you what a special friend you have become to me over the last few months, and I am so happy that we got to know each other so well. May God bless you and keep you safe in His hands, sending you my love, Philip x"

Sinead visited us on October 27, and at the same time Dad's youngest sister Gabrielle was home too with us. When I met Sinead for the first time, she told me that they had heard so much about me and that having contact with me meant so much to her mother, that it helped her to take her mind off her illness. I asked her if she had received the letter, she told me that she did, I was so happy and relieved to hear this. Sinead had to cut her trip to Ireland short and return home with her partner and children on Monday the 30th because Edith's condition had deteriorated further.

On November 1, Edith died.

I made the decision that I wanted to go over to England for her funeral. My parents agreed with me, that if anyone should be going, it should be me as I was the person who contacted her. Dad's sister, Veronica, was going over for her funeral so we both travelled together to London. We stayed with Gabrielle, and Dad's older sister, Angela, came down from Birmingham. We all travelled together down to Southend, where Sean and Edith had lived. Their home was in a seaside town which was just outside the city of Southend. As we approached the seaside town, I felt a little anxious, because I was about to meet my uncle Sean for the first time ever and his family also. I felt like such an outsider there, nobody knew me except for my other aunts. We drove down the street where they had lived and when we parked, I was greeted by my aunt

Molly and her husband, John, who were both delighted to see me. I was standing alone with them as my other aunts had gone up to the house. I got nervous and asked Molly would she take me up to meet everyone, because I did not know anyone here apart from Sinead whom I met for the first time in less than two weeks.

I walked up with Molly and John, and I approached a tall thin and grey-haired man in his early seventies. It was Uncle Sean. It was uncanny how much he looked like my father. I introduced myself to him and he gave me a warm welcome and thanked me for coming. We spoke for a few minutes before I was met by Sinead, who thanked me so much for coming. She told me that she knew I would be here, and that out of everyone here today, it was me who her mother would have been so happy to see. I had some tears in my eyes when she told me this, she then took me around to meet everyone else. I met her brother Christopher and her two sisters Debbie and Siobhan. They told me they heard so much about me and were so happy to see me at the funeral.

After we spoke for a few minutes, a hearse pulled up outside the house carrying inside a coffin with a wreath on top. Everyone went silent and walked outside.

It was time to accompany Edith on her final journey. I passed the hearse and all I could say to myself was, "This is the closet we will ever get to meeting each other. Why did it have to be for this occasion?"

I sat in a row with my two aunts, Molly and Angela as the service started. I could not take my eyes off Edith's coffin as she lay there. This sad occasion was the closest I got to meeting her.

As we exited the crematorium after the service, I went over to console Gabrielle. She was so upset. To say goodbye to someone who was part of your family for almost fifty years can never be easy. We then went to Edith's church for a memorial service.

We entered the packed church and took our seats at the top. It was a beautiful service to remember a beautiful lady. Chris spoke about his mother as did two of her friends. Such lovely things were said about her that it made me feel great, I really got to know this amazing lady more over the course of the service. At the end everyone joined in to sing and clap to the hymn *Lord of the Dance*.

After some refreshments in the Parish Hall, we returned to Chris' house for a few drinks and some food. We sat in his garden surrounded by heaters. I sat with Sean and my aunts as they talked about the old days.

This allowed me to chat more with Sean and gave me the opportunity to get to know him. I invited him to come to Ireland in the following May to celebrate Dad's 60th birthday. I had been planning a surprise party for Dad and I was hoping to get all of Dad's sisters and brother home to celebrate it. It would be more of a reunion as they had not all been together for so many years. In fact, I do not think there was ever an occasion where all eight siblings were together at any point in their lives. I was hoping to have the six of them home. Sean told me that he would think about it.

We returned to Gabrielle's home around 9 pm because she did not want to be driving so late at night. That evening, I left West-Cliff with a new family connection, an uncle, and cousins whom I never knew before. I was happy to have come here and meet such lovely people, but I wish that it were under different circumstances.

I landed back in Ireland two days later, and I had one thing on my mind, get Dad and Sean on the phone with each other.

After a few days, I got Dad on the phone to Sean. He was so happy to speak with him. Even though it had been fourteen years since they spoke, once they started chatting, it was as if time never mattered.

It is not easy to do something like this, especially when time has passed, and the relationships grow distant. I knew this would be the beginning of a new and wonderful relationship, not just for Dad and Sean, but for the whole family.

That is the thing, sometimes we are afraid of change, that we simply stay the way we have always been, stuck in our comfort zones, where nothing can interfere or change.

The months had passed by so quickly that we found ourselves preparing for Dad's birthday. We had decided to tell him that we could not keep it a secret. We were having a big family get together and it would be impossible to keep it from him, especially with two of his sisters staying at ours!

Among the family members coming were Uncle Sean and his eldest granddaughter. I was delighted when we got the word that they were coming home for this. This would be the first time in fifteen years that they had met, and I was excited.

They were staying down at Veronica's for the few days for which they were home for. Dad worked a few days down there with her, so he met them the day they arrived. He was delighted and could not wait to tell us when he got home. The next day, Dad and I went down to bring him up to Callan for the day. I was greeted with a big smile and a firm handshake when I saw Uncle Sean sitting there in the kitchen. We spent the whole day together taking him around to different places, his parent's grave, the family home in Callan in which they lived and finally we took him to Killaloe, the home in which he grew up in when he returned to Ireland when he was very young, although it was in ruins, he still could picture everything the way it was back in his youth.

We then took him to a few other scenic spots around the town. Both him and Dad were constantly talking, joking, and laughing. I was so happy that they were together again, the way brothers should be. After the party, I spoke to Sean and told him that I was planning on visiting London in July that year and I would like to visit him. He was delighted and said I would be very welcome to stay with him. Sean made a few more trips home to Ireland over the years but I guess the older we get, the more we want to stay at home and keep our feet on the ground. Almost every year since then, I have made the journey over to spend time with Uncle Sean. Sitting in his smoke shed and drinking tea is my favourite thing to do. My parents have made it over to stay with him on numerous occasions over the years. Today, as I live in China, no trip home to Ireland is complete without seeing him. This is something that I look forward to every single year.

Looking back on this memory, it is by far the greatest memory, also the greatest achievement I hold dear. I reunited my family, brought two brothers together who were once strangers for the best part of fifty years. Now we share memories, laughter, and phone calls. Sometimes, we must lose something or someone to gain more. Words cannot express how proud I feel about how close my father and my uncle are now.

With Uncle Sean, 2018

Chapter Three

Losing My Travel Virginity

Up until 2012, I had enjoyed 'safe travel'. Personally, I define safe travel as 'going somewhere on a trip to visit someone you know or somewhere you know, like visiting a friend in Australia or visiting family in London'. I define these two examples as safe travel, because I knew who I was going to visit, I knew what was going to happen on the trip. In one sense, it is pre-determined.

I knew exactly what I was going to do and what I was not going to do. This in one case limits your travel experience because you know what is going on, the chances of spontaneity can be rather low.

Going back to chapter one when I mentioned that I went to New York in 2012, I am going to share with you my experience of travelling somewhere solely alone and completely in control of everything I did.

After graduating from university in May 2012, and working part-time in the gas station, where I worked since 2006, I realised that it was not enough for me. I was working less than twenty hours a week and it would be better for the company to bring in someone with less experience to cover shifts simply because it was cheaper labour to have them and not me. So, working such a few hours was not enough. To make things worse, I had a loan on my car to pay off.

After my final exams, I took a job selling house alarm systems door to door. Direct marketing, for some people it is the best job in the world and to others it can be the most miserable job in the world. For me, it was a mixture of both opinions. The commission was great and when I sold alarms I benefited greatly, but if I did not sell, I did not get anything. Working in direct marketing often means you work solely on commission with no basic wage.

Unfortunately, I fell into the trap of thinking I could make a sale every day and meet my weekly sales target of four or five systems, I was wrong. I spent many evenings of the summer of 2012 walking in the rain, soaked to the skin, wearing a suit, and nothing to show for it.

Tip for the reader about this kind of work, there was one day in which I forgot my identification badge which meant I was not allowed to sell that day, because my badge would be needed for verification and to finish the sale. Now, my home was located ten minutes from the office and when I asked my leader, "Could I drive home quickly and get my badge?" She said "No", because there was not enough time, and I was going to learn not to forget my badge ever again, so I had to spend the whole day with her and some of the team, watching them sell and hitting both double and triple sales, while I sat like a fucking idiot in the car.

Now, I knew that it was my mistake that I forgot the card, but I am not sure if learning the hard way was the best even when I was able to get the card without causing any delays to the rest of the team (but by the time we left the office, I would have been back with the badge with plenty of time). For anyone thinking of going into a commission-only based job, think carefully.

One afternoon, before I went out selling alarms, I was checking my emails, I opened an email which was promoting a concert in New York. Barbra Streisand was about to play two dates in her hometown of Brooklyn, at the new Barclays Centre stadium. This would be the first time in her career that she would be singing in her native Brooklyn.

Now, I had previously seen her sing live in Ireland when she toured Europe back in 2007, in which she performed a concert in County Kildare. It was fantastic and it was the first concert that I ever attended. So, the fact that she was singing again in public and in her hometown was regarded as something I could not afford to miss out on. So, after checking the price of the tickets which were not cheap, I bought one.

Anyone who knows her music and her career knows that spending an evening in her company comes at an extremely high price! I thought about whether I should go, but then I realised that I was making great money and the price of the ticket would not cost me a thought. I remembered from my services marketing module in which I learned about tangible and intangible experiences and Barbra Streisand singing live was an intangible experience, so I booked it! A few minutes later I got an email informing me that I would be attending Barbra Streisand's 'Back to Brooklyn' show. One thing money cannot buy is experience. This was going to be an experience that I would remember for the rest of my life and that no amount of money could buy what I was going to feel, see and sense.

The next few weeks at work were difficult as I was not hitting sales and I could not sustain working in a commission only job. So, after spending a few days thinking about my job and my future, I left the job.

I was relieved, but at the same time, I had an expensive trip to New York coming up, which I started to regret booking.

My parents told me that I was still going and that they would help me out if I needed any money. I was grateful for them and my manager at the gas station gave me more hours. He had learned that giving less experienced workers more hours resulted in poor quality work and half completed jobs.

So, before I knew it, I had saved up enough money and I was ready to go to New York.

Unfortunately, I had devastating news one week prior to my trip. My best friend from university had died suddenly.

At the time of his death, we did not speak because we had a serious falling out two years prior and I always tried to make amends but sadly it was not meant to be. I was overcome with regret and grief and even I contemplated not going on the trip because I felt a bit unstable and deeply shaken by what had happened. Then, I thought that perhaps this trip could be what I needed the most, a distraction, a break that would help me accept and come to terms with what happened. Life is short. Too short, to hold onto hate or anger.

People should never fall out over small things, because not everyone gets the chance to make it right. Sadly, what happened between us two can never be made right. I must live with it.

Before I knew it, I was on my way to New York. I arrived extremely late at night and took a taxi to my hotel in Brooklyn. The taxi driver was very friendly and informative. He sensed my excitement about being here, so he invited me to sit in the front with him and gave me some information about different places that we passed on the journey to the hotel. This was my first time ever in America and I was already amazed by the size of the towers and skyscrapers.

As we arrived at the hotel in Brooklyn, I paid for the taxi and went inside. After checking in, I felt a tap on my shoulder. It was the taxi driver, he asked me if I had lost anything.

I checked my pockets and said no I did not. He asked me if I had my phone. I told him no and he smiled at me and handed me my phone. An extremely honest man.

The next morning, I woke quite early as I wanted to get out and explore New York. I went down to the dining room for breakfast, and after eating an egg and a yogurt, I embarked on my journey to the Statue of Liberty and Ellis Island. As I stood on the platform of the Brooklyn subway, I could feel the heat and the smell coming from the subway tunnels, it was so warm that my back was sweating from the steam.

Upon arrival to Staten Island, I could see Lady Liberty holding her torch on the island and across from her was Ellis Island, the first point of contact those thousands of immigrants had with America. I thought of the hundreds and thousands of Irish people who made the journey to America during the famine in Ireland in the 1800s. When I thought of that perilous journey they made for a better life, it made me realise that us Irish, a nation of immigrants is no different from the thousands of people, who still go to better lands in search for life and meaning, today.

As I arrived at the ticketing desk for the two islands, I noticed that there was no crowd in sight and thought it would be extremely easy to access them. After I purchased my tickets, I turned the corner only to see a queue about a mile long! Even at 9 am in the morning, there were hundreds, if not thousands of people waiting in line to visit the islands. I immediately regretted coming but then I read my ticket for Liberty Island, and it said it was valid for three days after purchase, which meant I could come back another day and try get there earlier to avoid the long lines. That is exactly what I did, so I turned back and went off exploring.

I spent the day shopping in stores such as Macy's and Army Navy, and taking in some sightseeing, which included a visit to Wall Street and Ground Zero. In the late afternoon, I decided that I wanted to go to the arena and collect my concert ticket to avoid the long delays the next day. Luckily, I had my passport with me, and I went to the Barclays Centre station where I was greeted with huge posters and screens with Barbra's face all over them. There was a bit of a crowd gathering as some were trying to purchase tickets and some people were complaining about their seating, that their seats were too far from the stage and the prices for them were scandalous.

Upon entering the arena, I was screened by security, and I was able to collect my ticket without any problem. I handed in my passport along with my booking letter and within a minute I was holding my ticket for the show. I then returned to my hotel and relaxed for the night.

The next morning, I got up at 6 am, skipped breakfast and made my way to Liberty and Ellis Island to avoid the long crowds that I witnessed the previous morning. Even though I got there around 7 am, there were still lines of people. Thankfully, they were not like the previous day, I decided to bear the long wait in the cold and visit the islands, because I would not be having the time to visit them for the rest of the trip.

After about an hour and a half waiting in line, I boarded the ferry that took me to Liberty Island. The cold air made my fingers numb, and it was hard to press the buttons on my camera to take photos of the statue.

Upon arrival at the island, I was given an audio set in which I could listen to the history of the island and the construction of the statue. I spent about an hour exploring the island and taking many photos of Lady Liberty. I was in awe looking at her.

I then took a boat over to Ellis Island, where I learned all about the immigrants who passed through here. I wandered around as I saw people trying to trace their ancestors back hundreds of years. I began to wonder if I had any relatives on my mother's side of the family who passed through here many years ago. Unfortunately, I didn't have anything to go on at the time so I couldn't research (when I returned home, my grand aunt informed me that we did have family members who passed through, shame I didn't know this before I went so that I could have found out something).

I spent another hour or so wandering around Ellis Island. It was around 1 pm now and I wanted to get some food and get ready for the show that night. After a quick subway sandwich, I took a nap for an hour. Around 4pm, I showered and got ready for the show. I brought my passport with me just in case the security was screening people before entry.

As I arrived out of the train terminal, I was met by large crowds of people. As I stood there taking it all in, I looked to my left and saw cameras gathering and snapping videos and photos of a person emerging out of the train terminal. It was Streisand's son, Jason Gould, I was just feet away from him.

As I queued to enter the arena, everyone was pre-screened with detectors. We were reminded several times that using cameras during the performance was strictly prohibited and people would be removed if they were caught using recording devices. Upon entering and going through another security checkpoint, I then ascended several escalators to take my seat. I had a great view of the stage, even though I was at quite a distance from it, I could see everything clearly.

Slowly, the arena filled up and the seats around me began to fill too. I started chatting with some people who were incredibly surprised to hear that I had come all the way from Ireland to be here at this event.

Beside me were two people, friends, who travelled a few hours to get to the show, we talked for about an hour as the arena filled up. I felt out of my comfort zone with others here as most people were in their late forties up into their seventies. I guess I was one of the youngest fans here. But that is one thing I love, the fact that not just a music event, but shared interests can gather people together and let people be bonded by their interests and tastes. Age has nothing to do with it.

At around 8 pm, there was an announcement that the show would start soon. After an eight-minute overture performed by a fifty plus piece orchestra, the lights went out and following a fanfare, the lights came back on and there she stood. Wearing a black dress suit and jewels, Barbra Streisand was with us, smiling and waving to the audience as they applauded and welcomed her home to Brooklyn. I guess nothing sounds better than your hometown. "Who says you can't go home again?" she said as she finished her first song. To the audience's delight, she was indeed home again.

After a few minutes, she began singing with a voice that has echoed for over half a century, and a voice that like a good wine, got better with age. She chatted and laughed with the audience and the show lasted for over two hours.

She ended the show with a signature song, *Happy Days Are Here Again*, as Brooklyn's favourite daughter left us, the audience burst into applause and tears of appreciation and love.

Upon leaving the arena, I had the opportunity to give an interview about the show. I was delighted to be given the opportunity especially when I told them that I travelled from Ireland. The camera crew told me that Barbra would see these interviews and that they may be included in a future television special.

The next morning, I woke early, still reeling from the previous night. Reliving the experience by going through all the merchandise I bought and telling my parents all about the show. They were delighted for me and were so happy that I was safe and enjoying every minute of the trip. Later that morning, I visited the Empire State Building, and I was able to get to the 83rd floor, the observation deck. It was here I was able to view some beautiful panoramic views of New York, taking in Liberty and Ellis Island and the Brooklyn Bridge.

From there I visited Times Square and Madison Square Garden. Now Madison Square Garden played a huge role in my childhood and teenage years, as it is the home of wrestling. When I was a kid, I was a huge fan of WWE and visiting here brought me back to my youth. Even though I had grown out of the wrestling phase, it was still great to visit the place that is symbolic of the sport, and my childhood.

Short on time, I made a brief visit to Central Park, Trump Tower, The Rockefeller Centre, and Saint Patrick's Cathedral.

That evening while I was preparing to go home and pack my bag, I realised that I had bought so much stuff that I had to go back out and buy a brand-new suitcase which could hold everything I had purchased. After a few attempts, I was able to successfully finish packing my case for the trip home.

When I got to the airport, while checking in my case, I was informed that it was overweight. When I asked by how much, I was told it was overweight by 5kg! I had to pay $100 for the excess weight. But considering how much I had in the bag, between branded clothes, souvenirs, and chocolates, $100 was a bargain!

I arrived home and after unpacking my case, I reflected on what I experienced over the last few days. I think all of us who come home after a fantastic trip or even a concert feel that it is over, and what do we have to look forward to next? I guess we could call this 'P.H.D.', post-holiday depression! I have suffered from it many times and I do believe it is real. For we spend so much time planning a trip and waiting for it, that when it comes and goes, we have an empty feeling and we ask ourselves, "What do we have to look forward to next?" The best remedy for this is to start looking at the next adventure!

But to summarise what my New York adventure gave me, it gave me the confidence and the courage to travel alone, and to explore a place by myself without any worry or anxiety. I experienced so much there and in addition, saw one of the greatest singers of all time perform in her hometown. Travelling alone also helped to push me so far out of my comfort zone in terms of having complete control over my experiences. I decided what time I get up, where I want to eat and what I want to experience.

Sometimes, when we travel with others, we can have opposite ideas and plans of what we want to do and where we want to go. So, sometimes solo travelling is necessary for us, so that we can do what we want and when we want. I would not describe it as being selfish, I would describe it as doing what we want. Having total control over our travel expectations and ideas is healthy for us. It also means that we do not worry about being alone, because when we travel somewhere, the chances of meeting new friends are exceedingly high.

For when we do meet people, we meet people of similar mindsets and attitudes, and these people can teach you so much about life. From my experiences over the years, some of the most influencing people I have met have been through travelling. Just take my Jordan trip as an example and what I got from it.

Finally, losing my travel virginity awoke something in me, an itch or as some people describe, the travel bug. The travel bug can be a contagious infection and it is something that is only treated by travelling more and I am proud to say I am indeed highly infected by the travel bug, and happy to bear it until the day I die!

Chapter Four

What Brought Me to Now

I will be forever grateful for this trip and most grateful for the people I met here, for I do not think I would be the person I am today, nor would I be where I am today if our paths never crossed.

It is always best to start a story from the beginning. So, that is exactly what I am going to do. I will tell you about the life changing experience I had while exploring Jordan.

I often reflect that if I did not go on that trip, I would not be where I am right now, be doing what I am doing or writing this book. Even more, I would not have visited all the amazing places such as Cambodia, Israel, Palestine, Malaysia, South Korea, Borneo, Singapore, and Vietnam, and not to mention have had my fantastic life in China, that I am profoundly grateful for.

Going back to January 2015, I was deciding where to go for a holiday. I was torn between a few different places, which included Morocco, Jordan, and Israel. I considered Morocco, but I felt that it was not far away enough for me, and I knew a lot of people who have visited there.

Then Jordan. I was fascinated by the culture here and the ancient history of this land. I had always been fascinated with Petra, the ancient city carved into the mountains of Jordan and not to mention *Indiana Jones and the Last Crusade,* where in Petra, the final part of the movie was shot.

Finally, Israel was my ultimate travel destination and a dream since I was young. I was obsessed with Jerusalem, the culture, the fact that it was the birthplace of the three major religions. Unfortunately, in the Summer of 2014, there was unrest again in the West Bank and it seemed that my visit to Israel had to be postponed, due to the conflict there.

After a long time thinking, I decided to go to Jordan on an eight-day group tour. I booked two weeks off with my area boss who was reluctant to let any of us take any holiday time. As this would be a once in a lifetime experience, I spent the months prior preparing for my trip, as well as preparing my visa and vaccinations. I also had to put up with an unmerciful amount of shit from my work, which saw me moving store and I also found myself suffering from anxiety and depression. I received phone calls from my area manager shouting at me and threatening to demote me. It got to the point where my heart would beat rapidly when his phone number came up on the phone's screen.

This was about the time in which I started to suffer from emotional eating (I found myself craving certain foods like hot food with a large quantity of salt: I would put six sachets of salt on a hot dog, and after I ate it, I would lick the salt from the foil), and I was getting very irritated by so many things.

Before I knew it, the time was quickly approaching and my trip to Jordan was almost upon me. With the stress of my current job, I began to think about some other options. I considered becoming a personal trainer or becoming a foreign language teacher. Right before I left for

Jordan, I conducted a lot of research on TEFL courses and even made inquiries with a company that offered them. I told the lady on the phone that I was interested in the Middle East as a destination and that after my upcoming trip to Jordan, I would decide about whether to go ahead with studying a TEFL course.

The day before I left for Jordan, I had a huge argument with my store manager which resulted in me crying and broken. Enough was enough, I basically was halfway with my decision to go ahead with the teaching course (I also wrote my resignation letter that day and I was ready to go back down to the store and throw it on the table to her, but my mother calmed me down and told me to wait). So, as you can see, I was not exactly in the holiday spirit as I had hoped to be. The next morning, my brother dropped me at the airport, and I boarded a plane first to Frankfurt. Soon, I was going to be away from my work and the stress and anxiety that it had brought me. I had a different sim card in my phone so that I could not be contacted or bothered by anyone at work.

As I sat in the airport in Frankfurt, waiting for my connecting flight, I started thinking of what was said to me by people about my trip to Jordan. I was told that I was stupid, crazy, and had a death wish to go to Jordan. Even though, Jordan is close to Iraq and Syria (where there was war and fighting) I never considered that I was endangering myself by going. I would respond to this negativity by saying that I could walk down the street in Kilkenny and get stabbed. So, my trip to Jordan was not really met with positive and happy wishes, but I did not care. I was going, and nothing and nobody was going to stop me.

I arrived at Jordan's Amman International Airport after midnight, and I had prearranged with my hotel for a taxi to collect me and bring me to my hotel. A few minutes into the taxi ride, the driver went off the main road and started driving faster. I immediately thought I was being kidnapped! He then slammed on the brakes and looked at me, smiled and said, "We need to change the car!"

He asked me was I scared with the way he drove, to which I replied, "Just a little!" and I told him I thought I was being kidnapped, with that he started laughing. His name was Shaukat, and he was the nicest taxi driver I ever had, for the whole duration of the drive, we chatted and talked about Jordan. We even arranged for the following day for him to meet me in the morning and take me out to explore Amman. I agreed to this because my group tour would not commence until late the next evening, so it was a great opportunity for me to get out and explore the city.

I was to share a room with another tour member but when I got to my room, he had not arrived yet. So, I had the room to myself. I was a little apprehensive about sharing a room with a stranger because it was something that I had never done before, but it was going to be another comfort zone that I would break out of.

I thought that experiencing this group tour would introduce me to some great people (you will find out just how right I was about this later in this chapter). The next morning, I woke incredibly early. I think I just had a few hours' sleep. The anticipation and excitement were too much for me and I just could not rest properly. I was up early and ready to go. Shaukat arrived around 9:30 am and we went off to explore. He asked me if I had any questions about anything, and I saw this as an opportunity to ask him some questions. I was very eager to find out why Muslims do not eat pork. I asked him and I loved his response: "Who would eat an animal that eats its own shit?" A good response. I also asked him about a place that was of special interest to me. I am an avid fan of the video game series '*Uncharted*' and in one of the games, Nathan Drake goes in search for the lost city of Ubar, also known as 'Iram of the Pillars'.

This game series has provided me with so much inspiration that it inspired me to go to Borneo and to Jordan. Playing the video games took me to places such as Shangri La, in the Tibetan mountains, the jungles of Borneo and Iram of the Pillars in the empty quarter known as the Rub'l Kali Desert in Yemen and Saudi Arabia. These games helped to spark my passion for travelling exploration. Even though the legends and places are real, they are portrayed and enhanced as fictitious stories and cities in the games. There are certainly some truths to their existence. I had done research about this place prior to visiting Jordan and I asked Shawkat about the place. He told me that this was a real place and that it is mentioned in the Qur'an, and that I would see the Seven Pillars of Wisdom located in the Wadi Rum Desert which were documented by Lawrence of Arabia. As you can see this trip was going to be something special and meaningful for me.

The first stop on my day tour was to the Amman Citadel. The ancient citadel has been inhabited by many great civilisations throughout the course of history and it was the very first time I really experienced seeing an important ancient archaeological site. The most impressive and important site is the Temple of Hercules, from the Roman era. Here you can see the pillars and among the stone ruins you can see the remains of a 39ft statue of Hercules, but unfortunately all that remains are fragments of an elbow and his fingers.

Another point to mention, I encountered the squat toilet for the very first time at this site!

You all will be too familiar with these as I discuss them throughout the book and hence the title of the book!

The next stop on the tour was downtown Amman. We parked the car up and strolled down the markets. I must admit it felt strange being the only westerner in the sea of people who filled the streets of Amman. The markets were fantastic with such delicious fruits and vegetables. I could smell the freshness of the food and the spices as I walked through and listened to the conversations of the locals with the market vendors. I really was in the middle of a different culture.

Suddenly, there was a voice talking over large speakers around the city. As it was Friday, it was time for the people to go to the mosque. This was the call to prayer. Some people went to the local mosque, which was nearby, while others just dropped to their knees from where they were standing and began praying. This was a defining moment for me on my trip, as here I was witnessing something amazing. Just to watch them pray and worship and to be standing in the middle of them watching with amazement and curiosity. This was my first ever cultural experience, and I loved it. My guide told me that I could walk up to the gates of the mosque and look inside, but I was not allowed to enter it. I was quite happy to do this as I was very curious about mosques and the Islamic religion.

Following a further wander around the markets where I enjoyed some local fruits and dates, we went to eat a traditional lunch, falafel, and hummus. Now, I did not know what to expect in terms of food here in Jordan, but I had previously eaten hummus and I had developed a taste for it. We enjoyed a huge platter of falafel and hummus. The restaurant we dined at is the most famous restaurant in Amman and the King of Jordan is known to come here and enjoy the delicious food that it offers.

After eating my weight in falafel and hummus, we embarked to the Jordan Museum where I explored for about an hour by myself. Here, I saw so many artefacts from Jordan and its ancient past, and something of huge interest to me. Within these walls of the Jordan Museum lies the Copper Scroll, which was found with the Dead Sea Scrolls. Unlike the Dead Sea Scrolls which were written on Papyrus, the Copper Scroll, like its name implies, was written on copper. This scroll in fact is a map to an ancient treasure which is buried in Jerusalem somewhere. I was very intrigued about this because I had watched some documentaries regarding this artefact, so to see it in front of me, with my own eyes was something else.

Another thing that I learned not just on this trip but on other trips too, that we always see places and things on postcards, on television, and in books, but when we see them in person, when we are standing next to them, it brings a whole different meaning to the word experience. We see them with our own eyes, not through the eyes of another person, or their point of view, also not even through their camera. We see it for ourselves, we create our own feelings solely based on our vision. Something that can be difficult to do is to write about a place or a thing that you saw or experienced, because at times, the words cannot do justice to it. Therefore, you simply must go and see it for yourself and just let the feelings go through you.

The remainder of the afternoon was spent at the Royal Automobile Museum, where the King of Jordan's very impressive car collection is on show. Everything was there, from Rolls Royce to models used in movies such as Transformers, state cars, a pope mobile used by John Paul II on his visit and the car used by Queen Elizabeth II and Prince Philip while they were here during a state visit. For an auto fanatic, this was heaven. After a fantastic day exploring Amman with Shaukat, he dropped me back to my hotel and it was time to meet my group that I would be sharing the experience with for the next eight days.

I was not in my room for long when the door opened, and my roommate walked in. Merv was from Australia and had just finished two years in London working with a law firm. He had been travelling for a few weeks at this point and he was slowly making his way back to Australia, but not before taking in many countries along the way. We clicked instantly. Talking about travelling and other topics, we became friends within a short time. After about an hour chatting, there was a knock at the door and Craig walked in. Craig was a 63-year-old from Canada who had just finished two weeks travelling around Egypt. He came to our room to introduce himself, which I thought was so nice of him to do.

The three of us walked down to the conference room where we met the rest of our tour group and our CEO (Chief Experience Officer) Zuhair.

In total, there were twelve of us on the tour. From Ireland, England, Australia, Canada, and America, we had a nice cultural mix on the tour. After Zuhair gave us the tour itinerary in detail, we went down the street for some food. On the way down, we all mingled with one another about where we were from and talked about travelling. It was awesome to be around such cool and well-travelled people. I chatted a lot with Tom and

Jade who were from England, Amanda who came from America, Stephanie, Annie, Aaron, Paul and Amber from Australia and Thomas who was from Canada. What was great was that each of us was able to connect with one another on some point or on some common interest. It was wonderful. I had such a great feeling about this trip, and I was extremely excited to embark on it, as was everyone else in the group.

The next morning, after breakfast, we embarked on our first stop which was to the historic site of Jerash. It was located around one hour away from Amman. On the way, we learned so much about Jordan and its long and fascinating history. Jerash was a Roman temple to the Greek Goddess Artemis and the ruins of the great temple are remarkable, they include auditoriums and streets. After this visit, we travelled to the Dead Sea, the lowest point on earth. We learned about the unique history of this area and its significance between both Jordan and Israel. For it is believed that this area was the site of the sin cities of Sodom and Gomorrah. As I stood there looking at the Dead Sea, I looked across the sea and my eyes met the land known as Israel. I thought to myself, '*I am so close to it now, but one day soon I will step foot there.*' Zuhair told me to look over to the right and you can see some buildings, he informed me that this town was Jericho. It is crazy when you can be so close to somewhere you really want to experience but you must wait, and that can be difficult to do.

The Dead Sea was amazing. We floated in it for hours and covered ourselves in the healing mud. We spent the entire day there. We also all took the traditional photo of lying in the water, floating (because of the extremely high salt content) while reading a book. Throughout the course of the time spent here, we all really got to know each other better.

After we spent the day at the Dead Sea, we returned to Amman and to our hotel, where after a short time to refresh, we went back down for dinner and some shisha.

As a non-smoker (at the time), I quite enjoyed smoking shisha. It gave me a great feeling of relaxation. Come to think of it, we all quite enjoyed smoking it! The following day, we visited Mount Nebo which is where Moses saw the promised land (Israel). Again, the land in the distance of Mount Nebo was that of Israel. Along with the help of a map, we could see what cities lay in what direction. We were able to see towards Bethlehem and Nazareth, but by looking straight ahead on an extremely clear day, people could see Jerusalem. Again, this gave me time to reflect that I was so close to visiting the city of my dreams.

Following our visit to Mount Nebo, we went to Madaba, and visited the Greek Orthodox Church of Saint George. This church is famously known for housing the Madaba map of the Holy Land. The mosaic map dates from the sixth century A.D., and it highlights important holy places such as Jerusalem, Bethlehem and even has record of the sin cities of Sodom and Gomorrah. For lunch this day, we ate the most delicious falafel wraps. I think we ate two or three each. They were fantastic, and we could not stop eating them.

By now, some of us had established a little circle of friends. There was Merv, Amanda, Aaron, Steph, Tom, Jade, Amber, and me. This was special for me, as I was not really included in things growing up. I was usually left out. We would all hang out and do everything together. We became friendlier with each other as the days of the tour passed. That afternoon we spent exploring the crusader castle of Kerak which was built in the 1140s. If you have ever watched *The Kingdom of Heaven*, you would have seen it in the movie.

That late afternoon we began a three-hour bus journey to Petra. We spent the time getting to know each other better and asking each other some fun and probing questions. Upon arrival at Petra, we were quite tired after the long journey, but our guide Zuhair gave us the option to take part in the 'Petra at night' experience. This experience entails descending into the caverns of Petra in darkness, to see 'The Treasury' lit by candlelight and enjoy traditional music. Only Merv and I opted to do this and after a quick dinner, we embarked on our evening activity.

As night was falling quickly, we descended into the caverns of Petra walk through the Siq, walking down the long narrow and windy path. By the time we arrived near the Treasury, there was total darkness around us. As we entered the canyon which hit the Treasury, we saw the huge outline of the building carved into the stone. I was amazed. After reading about Petra and watching it on television for many years, here I was facing it.

The event lasted around two hours and it was quite enjoyable. We got to listen to traditional music and saw traditional dance, while the candles lit up the monument in the background.

We went back to the hotel around 10:30 pm and we were all told that we had to be up at 5 am the next morning. The reason for this was that Zuhair wanted us to experience Petra at the best time, and the best time was at the crack of dawn, because the tour buses would not be there yet. So, after an extremely early breakfast, we embarked on the descent into the canyons to the ancient city. It took around an hour to get down there. Zuhair gave us so much information about the city and the history, it was interesting to learn about this place. We saw the different parts of the canyon that were featured in *Indiana Jones* and *the Last Crusade,* which lead to some of us whistling the theme song as we walked down.

Just as we were about to walk through, I could see the Treasury, but this time it was in daylight so we could see it in all its glory. Petra is known as the rose city because it is carved into the pink sandstone cliffs. It is an entire city filled with buildings and tombs, but this was something that I did not know about until I got here. We only ever see the Treasury in movies or in photographs, so I was quite surprised to learn about there being an entire city here, in the canyons.

We were supposed to spend the entire day exploring the whole city, but the first hour was spent almost entirely at the Treasury. Zuhair was right, there was nobody else there, we could take photos in peace and quiet without being disturbed and having people standing in our way. It is amazing to think that this city was built well before the time of Christ, and it remained hidden from outsiders for hundreds of years until the 1800s. It was the centre of the Nabatean Kingdom. Its people, the Nabateans, came from North Arabia and South Levant. After some time, taking in the beauty and detail of the Treasury, we began exploring the rest of the canyons. We began ascending to a highpoint from where we would be able to look down on the Treasury offering the best photographs of it. As we began ascending, we all started singing songs and telling jokes, but the best part of the hike was when Steph began singing '*Wuthering Heights'* by Kate Bush at the top of her lungs! I had heard the song but could never put the name of the singer to it, or even understand the lyrics due to my hearing loss. To this day, whenever I hear this song, it brings back memories of that day in the canyons of Petra.

After about an hour of hiking and stopping at different buildings and viewing points we reached the summit of the cliff where we were able to sit on the edge and look down upon the huge monument below.

It was here at this point exactly, staring down at the Treasury, that I had my awakening. My life changing moment.

While I was seated there looking down, many thoughts and questions entered my mind:

"What was I doing with my life?"

"You need to do more of this, because this is what you love deep inside."

"It is time to say fuck off to the 9–5 job, you need more from life."

"You need more goals than just hitting KPIs for a job and manager that would replace you in the morning for not hitting them."

"Why settle for just a piece of sky?"

The time for a change was coming fast and I had to be prepared to pursue that change.

After some time here, Zuhair told us that we needed to move down because the temperatures were reaching a scorching level of 50 degrees Celsius (122 Fahrenheit). This was starting to get too much for some of the group members and they decided not to continue after lunch. They decided to opt out of the hike to the other side of the canyons to see the Monastery, which is like the Treasury, but much bigger.

After resting for about an hour, the crowds were beginning to get more and more, so we decided that we needed to move onwards. Zuhair did not go with us, he left us off on our own to explore. I think about three of the women did not continue so they went back to the hotel and relaxed. The heat and the hiking had got to them, and they needed to take time to recover. But for the rest of us, the fun was continuing, as we braved the scorching temperatures by putting on additional sunscreen and protection. I was concerned with my high factor cream whether it would be strong enough to take the temperatures. I had never experienced heat like this before in my life.

So, we began our ascent into the canyons again. We talked, laughed, joked, and played tricks on one another as we walked through. We helped each other when we found it difficult and when we needed to take a rest to recover. The hike up to the Monastery took well over an hour, maybe an hour and a half, but there were plenty of monuments and architectural delights to distract us from the heat during the exhausting hike. As we got closer to the Monastery, the heat was beginning to drop a little, but it still caught us at times, and I found myself layering on the sunscreen. I was wearing a keffiyeh (which I bought specially for this trip and very similar to the keffiyeh worn by Nathan Drake in the *Uncharted* game series), hat and glasses, along with long sleeved and long-legged hiking pants, but it was not enough to protect me. However, I was not stopped.

We walked through a market on our way up, it was run by some Bedouin people who lived in the city. We were pestered by them to buy trinkets and water. Some of our group members stopped to try and haggle but they didn't make much progress on this. I guess working at this trade, trying to scam tourists and extort money from them had really taught the local Bedouin people to be masters of their trade.

After more hiking, we finally made it to the Monastery. We spent some time appreciating it in all its glory. It was here that I called my mother and told her I was standing at the top of the city. She was delighted and amazed to hear this. Another thing that I realised up there was that I had just seen a Wonder of the World. The first one for me, I was so proud of myself and seeing such an important structure began smouldering feelings inside me that I should be, no, must be travelling more; this one vacation a year malarkey was simply not good enough. I looked around at my new friends and thought what wonderful lives they had, travelling, exploring, and living, counting countries and experiences while all that I was doing was stocktaking, invoicing, and taking refrigerator temperatures. I wanted more. I needed more.

So, after a while taking photos and staring in awe at the marvellous buildings, we began our descent. We were told that we needed to be out by 6 pm for dinner at 6:30 pm. The journey down proved to be more tiring than the journey up. We found ourselves taking more frequent and longer rest breaks. I also ran out of hydration salts which I used in my water. I had supplied some of the guys with them to keep them hydrated. Eventually, we made it back to the Treasury, where we all took some more selfies and took in the building one last time as we left the park. One thing that meant a lot to each of us was that we stuck together, even when some of us found it hard and wanted to quit the hike, we motivated and helped each other through it. For it is moments like this that we remember and build friendships on.

That night we enjoyed delicious local food and went to this very cool bar. This bar was in a cave, much like the caves we explored in the ancient city of Petra. We spent some time there, but we were all very tired and exhausted from the day's hike, so we did not spend too much time out. My feet were very sore by the time I got back to the hotel and when I took off my socks, I saw the amount of water blisters on my feet. I could not believe the amount that had accumulated from the hike. They were not painful but just annoying and I was happy that they did not prevent me from enjoying the trip. The next morning, I have to say I felt a little delusional when I woke up because I felt out of it especially when I went the bathroom to brush my teeth. Instead of using toothpaste I realised I began brushing my teeth with hair gel! Perhaps, I was too dehydrated from the previous day, I do not know but I could not stop laughing at my own stupidity.

After breakfast, we began our journey to the desert of Wadi Rum where we would meet with the Bedouin, drive across the desert in 4x4 s and camp overnight. As we left Petra, our guide showed us Aaron's Tomb nestled on a peak (Aaron was the brother of Moses). His tomb was commemorated with a mosque built over it and lies on the highest peak in the area at a height of 1,350 metres.

We now began the long journey towards the desert. This would be the longest drive of the trip, taking five hours to get to the destination. We kept ourselves entertained by asking questions and talking. I have to say on this trip I cannot recall anyone playing on their phones at any point. It was pure human contact and conversation. So different from now. We made a few pit-stops on the journey, and we also stopped at a supermarket to get some food and refreshments for our camping trip in the desert. Some of the group stocked up on beer and whiskey while I stocked up on juice. As we approached the edge of the desert, we had to change from the bus into 4x4s, and it was here that I saw the seven pillars of wisdom which serve also as the title of the autobiography by T E Lawrence of Arabia. I had waited so long to see these pillars, because I had heard of them many times before. It was here we learned about

Lawrence of Arabia and his presence here in Jordan which was met with different opinions.

We soon embarked on the drive through the desert. This was another highlight of the trip for me. Here, I was sitting in the back of a 4x4 being driven across the desert, it was one of those 'being alive' moments for me. We stopped at various points of interest to take photos and to learn about their importance. We were also informed that millions of years ago that this was a seabed, and we could see evidence of this in the various arches and the structures of the rocks.

Soon after we arrived at a Bedouin camp where we met the chief of the people and enjoyed listening to traditional Bedouin music and perhaps had the nicest tea I have ever tasted. It was a delicious, sweet Bedouin tea that they serve visitors, I had quite a few glasses of it. It was a shame that they did not have any for sale because I would have bought a few bottles of it. We learned about the life of the Bedouins and how they came to live in the desert.

Later that day, we hiked to take in a beautiful sunset over the desert of Wadi Rum. Although we had to wait quite a while for it to happen, the setting sun over the desert was magnificent to watch. It was total darkness when we arrived at the Bedouin campsite where we would be staying the night. The purpose-built campsite located in the shadow of a rock formation offered shelter from the sun to passing caravans.

The campsite was complete with toilets, dining area, fireside, and tents. That night each of us had our own tent to sleep in. Dinner was prepared at the base where we took the 4x4s. It consisted of chicken, beef, hummus, rice, and vegetables. Middle Eastern food is my favourite cuisine. After dinner, we gathered around the fire talking, laughing, telling jokes and stories. We enjoyed asking and answering some probing questions. It was nice to have people with the same dark sense of humour as myself.

That night we ate so many snacks including s'mores that Amanda made for us with bananas and chocolate. They were delicious. We spent the night by the fire and slept beside it. I must admit it was quite cold as there were no clouds in the sky, just pure sky. I never saw the night sky looking as wonderful as it did that night. The stars, the Milky Way and the zodiac constellations were naked to us. Some of us woke at 5:30 am as we were going to ride camels across the desert. This was another great moment for me. Unfortunately, soon after I embarked on the camel ride, my stomach began to feel sick. Perhaps, it was the food that I ate the previous night, but I was not letting it ruin my morning ride. I felt like a real adventurer crossing the desert on camel-back.

We took in a beautiful sunrise overlooking the arches and hills in the desert and about an hour later we arrived back at the base camp from where we began the trip the previous day. I was never so happy to see a toilet!

We were now on the last day of our trip before returning to Amman. After breakfast, we took a short bus ride to the port city of Aqaba, where after checking into the hotel we would enjoy the late morning and afternoon snorkeling in the Red Sea. We boarded the boat and left the port to go out to the snorkeling area. Here, we were swimming in the Red Sea bordered by Israel, Saudi Arabia, and Sinai. I could not believe where I was. As you know the Red Sea is of biblical importance as this is where Moses parted the sea to lead his people to the Promised Land.

I was not a fan of the water (still not), but I went in anyway wearing the snorkeling mask and the lifejacket. I saw perhaps the most beautiful marine life I have ever seen in my life. I was apprehensive about going into the water but soon broke that comfort zone, how happy I was to do this.

Back on the boat we relaxed for a few hours before going back to port. There was no alcohol on the boat as Zuhair and the captain explained to us that they were preparing for Ramadan, which was beginning the following week. Therefore, they were not carrying alcohol on the boat.

We returned to the hotel and had the remainder of the afternoon at our leisure. Some of us decided to go and chill at a bar. I was about to learn some devastating news. Back at the hotel I went scrolling through Facebook when I saw the breaking news:

"Sir Christopher Lee dies at 93." I immediately felt tears running down my cheeks as I read the details of his death. Merv asked me what was wrong, and I told him that Christopher Lee had been my hero, my idol and inspiration since I was a kid and now, he had died. I got upset over this. Anyone who knows me will know and tell you how important Christopher Lee is to me.

Whenever I felt down or sad, I would always watch one of his movies, whether it was *Dracula* or *Star Wars*. I always had his movies on standby to watch. It is crazy how we take things for granted and expect people to always be there, but one day they will have to go and sadly it was his time. I tried not to get upset while we went out to the bar however, it cheered me up to be with the group.

After relaxing at the bar for a few hours, we went back to the hotel and got ready to have dinner again. Many of the group members felt exhausted after the early morning and the snorkeling. So, towards the end of the night, only Zuhair, Aaron, Merv and I were left standing. We ended up on a rooftop bar having drinks and smoking shisha again. We talked into the early hours of the night about many things.

The next morning, we began the journey back to Amman where we had started the tour the previous week. We drove past The Dead Sea again, but this time we passed something of great interest. As we drove, the bus came to a stop, and we got off. Zuhair pointed us in the direction of a hill above us and standing on the hill was a rock formation that looked like woman.

Now the biblical story of Sodom and Gomorrah teaches us that when God told Lot to take his family away from here and not to turn around and look while He was destroying the city. His wife turned to watch the destruction and she was turned into a pillar of salt. That is the biblical story that we learned, although it was quite cool to see this and perhaps there is some truth to it, there remains the question of the authenticity of this landmark.

We then looked at The Dead Sea again and we could see by the rocks, just how much and how fast the water was disappearing. Zuhair told us that The Dead Sea was slowly disappearing over time.

While driving back on the bus, I was hit with reality extremely hard. This was the last day I would spend with some of the nicest and most wonderful people I had ever met in my life. I did not want it to end. I sat there with my headphones listening to music and the song *'Words'* by the Bee Gees played as the reality of going back to my soul-destroying job became clearer. Although, after having time to reflect on my life and my current situation on the mountain-top of Petra, I was determined to change my attitude and go back more motivated than when I arrived. Zuhair asked us to come up one by one and take the microphone and asked us to reflect on our Jordan adventure.

When it was my turn to go up, I said the following:

"You know when I told people where I was going, I was told I was crazy and stupid to want to go to Jordan. That why could not I be like everyone else and be satisfied with somewhere like Spain. I told them that I wanted to push myself out of my comfort zone and see something that everyone has never seen before in their lives, and to feel things they never felt before. This trip has done that for me, but perhaps the best part of this trip for me was meeting all you wonderful people. You do not know how much it meant to me to spend this time with you, you have inspired me more than you will ever know, and you have really shown me what it is like to travel the world, and for that I am eternally and truly grateful. I feel a powerful and positive change coming in my life."

I was applauded and told that people felt the same way about the connections I made with the group.

After arriving back in Amman, we left our luggage at the same hotel in which we started, and because I wasn't staying the night (I had a night flight back to Frankfurt), I left my backpack in Merv's room. Once again for the final time, we went out to a bar to enjoy the remaining time we had together. We met again for the last time and to have a final dinner. We presented Zuhair with a card and generous tip and told him that we appreciated everything that he did for us on the trip and that he really made our trip special. After some photographs, the time had come for us to go our separate ways.

It was hard for me to say goodbye to some of the group, especially Merv, Amanda, Aaron, Tom, Jade, and Amber. I felt that I really connected with these wonderful people during my time in Jordan and saying goodbye was hard. It was a mutual feeling between us. We had arrived as strangers and left as friends. We told each other that we would meet again at some point soon (to my surprise, little did I realise just how soon we would meet again).

It was not long before I found myself in Frankfurt and then back in Ireland. I was ready to embrace the changes that were about to come but they did not come easy. Before I returned to work, I took a few days reflecting on what I had experienced on that trip to Jordan. To be honest, I was feeling both great and sad. Great because it happened and sad because it ended. I did not know if I was fully prepared for making the changes. As someone who does not adapt to change so well, I had decided to push out of my comfort zone and embrace what would happen once I got the cogs turning.

As mentioned earlier in this chapter, I would let Jordan be the decision maker about whether I would engage in the TEFL course. The day after I got home, I rang the TEFL company in London and purchased the course. I paid for a 300-hour TEFL course over the phone with my bankcard. I said to the sales assistant that it was time to make a change and that I feel this would be the best thing I would ever do in my life, she agreed and said that I wouldn't regret it one bit.

My focus was set on teaching in the Middle East, and nowhere else. I knew there would be times when I wouldn't want to study it or be even bothered with it, but I had to be disciplined and focused on it, for this was my way out of retail, my one-way ticket out of hell, which is what my job became for the next five months of my life (you will read about this in the coming chapters).

I mentioned in the previous pages that I did not realise how soon we would meet again, here is how we met. Just a few weeks after the trip, Tom and his friend came to the Southwest of Ireland and together, the three of us climbed the highest peak in Ireland. That December, Amanda was touring some cities in Europe, and she included Ireland on her visit. I had the honour of showing her my hometown, introducing her to my family and exploring Dublin with her.

The following June (2016), Merv was over with his family in Limerick as his sister was marrying an Irish man. He took his cousins up to Kilkenny for a night and I was able to meet with him and catch up.

In 2018, I had Tom and Jade over with me in Guiyang, China where they spent some time teaching English in one of the schools I worked at. Most recently, on my last trip home in August 2018, I had Amber visiting Kilkenny during the same time I was there.

It shows you what can happen when you meet and connect with some wonderful people. Life is a mystery, and you never know where it will take you. For me life has taken both many good turns and bad turns, but ever since that trip to Jordan, the most turns I took ended up being good.

Now, I have told you about my journey through Jordan and what happened to me on it. It allowed me to see that there was more to life outside of my comfort zone. Everything changed because of that trip. My attitude, my mindset and just the way I saw everything.

The last thing I want to do is to tell you, the reader, that if you are not happy with your current situation, that if you do not like your job and where your life is going, you can change it and live an amazing life. You can guess from my stories that I felt like this, and that I have made the changes that are allowing me to live life as I wish. I have been to hell (on my own journey) and thankfully came up the other end better. I hope that my journey can inspire you and make you believe that you can do anything you want. I owe everything to that trip to Jordan.

Use those questions and thoughts I asked myself while staring at Petra. Here they are again for you.

"What are you doing with your life?"

"You need to do more of this, because this is what you love deep inside."

"It is time to say fuck off to the 9–5 job you have, you need more from life."

"You need more goals than just hitting KPIs for a job and manager that would replace you in the morning for not hitting them."

"Why settle for just a piece of sky?"

The Amman Citadel

Looking down upon The Treasury, Petra

The Monastery, Petra

The Seven Pillars of Wisdom, Wadi Rum Desert

The amazing group of people I met on the trip, The Treasury

Chapter Five

Chasing Rainbows

As previously mentioned in the Jordan chapter, I realised even before my trip that it was time for me to get a new job or even perhaps change my field of work. I had thought of two options. The first one was to go back to university and to study part-time to become a personal trainer. I had taken a huge interest in fitness in 2015 and I felt motivated that it could be an option.

The second option was to study to be a TEFL teacher (teach English as a foreign language). My friend had spent two years in South Korea doing this right after he finished university. As a matter of fact, it was in 2013 when I left for Australia, and he also left for South Korea. He had mentioned it to me even back then, even though I expressed some interest in teaching abroad, it had fallen on deaf ears as I was so focused and determined to try Australia. Nothing else mattered.

So, after Jordan, I knew exactly what I wanted to do. I realised that I would like to do something that would take me around the world and experience something different. I had decided that I would go for the teaching option. But I still was not completely convinced because I was worried if I had jumped straight into it, I might not like it.

So, I spoke to my friend, heard everything he had to say about his time in South Korea. I did some research of my own, and then I decided that this would be the path to take me to a new chapter in my life. But again, I still had little doubt. So, it was the trip to Jordan which was the breakthrough event that sealed the deal that would lead me to embark on my studies.

Upon coming back from Jordan, I was determined to do this. I booked the course, but hesitation due to problems at my job and a breakdown caused me to delay starting the course for a few weeks. I had got into a dispute with my manager over something simple, and it happened. My body and mind broke, I fell to the ground crying, not being able to speak for a few minutes. The only sound was that of me crying.

I realised that I had been broken by the stress and pressure which was brought on me from the job that I once had great enthusiasm for. As mentioned earlier the day before leaving for Jordan I had another dispute with my boss, after which I had written my resignation and had planned to throw it on her desk, but my mam had stopped me. So, I threw it into the glove compartment of my car.

But that day, the first day back from my trip, I had my breakdown, my boss told me to go outside and calm down. I went outside and spoke to my mam and told her. She said, "Do what you have to do." I took the letter of resignation and had it in my pocket as I returned to the office, but the mood was different. There was a huge sense of compassion and care from my manager, who had been talking to my area manager.

After a very meaningful and caring chat, I decided not to hand it in just yet. As a matter of fact, I told my boss that I had intended to study a course in teaching. She said it was a great idea. I must say, even though I rarely saw eye-to-eye with my manager at that time and we would fight like cat and dog, even though we had personality differences, she had my back and she cared. She knew the mental pain I was suffering from, which was caused by the job, and she really supported my decision to study to be a TEFL teacher.

For most of the Summer of 2015, I spent the time on my computer in my three-bedroom house which had belonged to my late grand-aunt. Looking back on it now, it took a lot of will-power and a strong mindset to spend at least one to two, even three hours a day staring at a computer screen, sacrificing, and passing up opportunities to socialise with friends, meet people and spend time with those dearest. But they understood my vision and my goal.

To be honest, Australia was something that I wanted to do, whereas teaching in the Middle East was something I needed to do. I must say, that looking back on the time I spent studying, I had an extraordinarily strong mindset and dedication to get it done. Not just getting my teaching course done, but reflecting over the years, I always had a strong mindset and I guess strong and clear goals about what I wanted to do.

Whether it was losing weight, getting my hypothyroidism disorder under control, or visiting the places I wanted to see, it was instilled on my mind. I often question where I got it from, but I always fail to see exactly where. I know I got some of it from my parents. I guess I always had a clear vision of everything, and I always remained hungry for more different things.

Now, getting back to my studies and the dedication I put into it. I guess I never even questioned myself in the hard times, "Is it worth it?", "Am I chasing a rainbow?" An idea that I could just walk into an Arab country and get a job.

I believed I could, and why couldn't I? If others were able to do it why the hell wasn't I? I just got on with it.

I must share with you my reason, my motivation for really getting immersed with my study. As mentioned earlier in this chapter, I had a very hard time in my job when I returned from Jordan, and that inevitably one person was responsible for giving me the drive, the passion, and the dedication to do my course.

One morning after my breakdown, I received a phone call from my regional manager at that time. I was at my parent's house in the country. He sounded so sympathetic saying that there is a lot of pressure in the retail industry, and it is not for everyone. But it is what was said next that pushed me even harder.

He said:

"Philip, you should think carefully about the decisions you will make in the next few days, because remember this, you are in retail for life. You will struggle to find any work in another industry, and you must get used to it and cannot blame personal trauma or stress for your mistakes."

I have italicised this and put it in bold font because I want you, the reader, to remember this, and please remember next time you have a quarrel with your manager over something that is not your fault. That YOU, yes you, have the power and the strength to turn your life into anything you wish it to be. Do not 'wait and see what happens' as some would say but rather what I say, "Go and do." You will certainly find results coming your way, your hopes, and your visions. If you just wait and see, many things will pass you by and you will not be able to grab on to it. Trying is always better than doing nothing, and if it does not work out? So what? At least you have tried something new. You had the courage to pursue something different.

There are two quotes that I want to share that really motivated me. They were spoken by two people who I have a lot of admiration and respect for. To summarise these quotes, firstly, they tell me that people who try put you down, who try to tell you that you cannot do something, they can push you harder to do what you want. Secondly, do not listen to the naysayers, prove them wrong.

American singer Barbra Streisand stated in 1999 during an interview:

"Actually, you have to be thankful for people like that, because in a way they gave you more courage, more energy, and more of a sense of your responsibility."

The late British Actor Sir Christopher Lee, who was one of my heroes and an inspiration to me said:

"He's too tall, much too foreign looking, he can't do comedy, he can't do a western. He can't do this; he can't do that."

"I've done it all! I proved them all wrong!"

So again, to summarise what I have just mentioned, you can do anything you want. Escape the prison of the dreaded 9–5 lifestyle, if you have visions and passions for something better go and do it otherwise, you may never know just how powerful and great you are.

Getting back to the studies. I had enrolled in a 300-hour teaching course which was quite compacted with the main 120-hour course and other courses, which provided me with specialised qualifications. By the end of October, start of November, I had completed the required 120-

hour course. I still had so much to do and with work piling on me, it was not easy to balance it.

At this time, I covered shops in neighbouring cities and counties before being moved back to my own city of Kilkenny, where I was put as manager of the small station there. I loved it and just three weeks into it, I suffered yet another breakdown. I had been covering holidays and sick leave in one of the Waterford stations for a few days in the week and at the same time managing my own store.

The realisation that I needed to seek professional help to deal with my anxiety issues came from a build up from a series of events, existing issues, a phone call in the morning to tell me that the fuel pumps on my forecourt were an environmental hazard, and then the worst, in the afternoon, another phone call to tell me that of my team members serving cigarettes to a young teenager, which was part of a health and safety sting which had serious consequences (this was also to be the first time I fired someone). All this happened on October 30, a Friday, I never forgot it and I do not think I can, this is a classic example of that line proverbial idiom, *"It never rains but it pours."*

The impact and backlash of this drove me to seek medical help from my doctor who diagnosed me with anxiety. I was a little happy with this diagnosis because I really felt stressed and anxious almost every day in my job.

I would jump with fear and my heart would start racing when certain mobile numbers came up on the phone. I knew exactly what was going to happen. It is with moments like this that you realise you need to seek professional help. I did. And it was the greatest thing I did. I went to a psychotherapist and I also found a personal trainer to teach me how to

71

weightlift. I discovered that exercise was a way for me to relax and to help my anxiety. I will always be grateful to my psychotherapist Dorothy and to my coach Ricky, for what they did for me during that time.

Getting back to my studies, as mentioned it was incredibly hard to balance managing a busy gas station and trying to study but again, I was so focused. I had developed a great routine every day. I finished work, went to the gym, got dinner, and settled into my studies for the night. I finally finished the course in March 2016, after receiving some job offers prior to it and being asked to take them in February. I wasn't prepared for that just yet because I had a world of things to do and to organise to prepare for my new career abroad. Now speaking of job offers, I had the most unique and interesting interviews, which led to an unexpected change of plan.

Chapter Six

A Change of Plan

My plan for my career as a TEFL teacher as previously mentioned, was to go to the Middle East. I was motivated and inspired that this was to be the destination where I wanted to pursue my career. So, I began to research teaching opportunities in the United Arab Emirates, Kuwait, Saudi Arabia, Bahrain, Oman, and Qatar. Even what inspired me more were the people I knew, who had relatives and friends in these countries teaching English and being told that there is a world of opportunities there for me. The nicest area manager I had, a lady, she spent a few years working in Qatar and I often chatted with her about her experience.

For teaching there though, to be honest, it was not that simple. There was quite a bit of red tape and many requirements for working as a teacher in the Middle East.

Firstly, you must respect the culture and traditions of the Middle East, depending on the country. This requires you to change your lifestyle, if you are a party animal and love to drink, rule out going to Saudi Arabia, because it is against the law there. You must follow the rules of these countries, otherwise you will not succeed there. It goes for everywhere.

Secondly, you must have teaching experience to ensure you even being considered for a position. Furthermore, some countries here also require a CELTA (Certificate in Teaching English to Speakers of Other Languages) or a DELTA (Diploma in Teaching English to Speakers of Other Languages). These are more intense and more widely recognised teaching certificates than an online TEFL course. As I was job hunting online, I was seeing many amazing jobs with some fantastic benefits and salaries, but the red tape that was 'no online TEFL certificates will be considered' made me feel little knocked.

Not uninspired, I decided to apply for three jobs at the same time. I applied for teaching positions in Saudi Arabia, Kuwait, and the United Arab Emirates. Three separate and interesting application processes that I will detail now. Then there was China.

Saudi Arabia

This job was for a university English teacher. The interviewer who was an American seemed keen on my application. We arranged a Skype interview at 3 am in the morning one Sunday. It went very well, but the only thing was that he told me to stop the TEFL course and start the CELTA course or enrol in an education course at a university for two years! I was not about to wait and commit to that. It would mean I would still be studying and preparing to go in 2018. I decided to scrap the idea about teaching in Saudi. I could not delay my plans.

Kuwait

This job was for a military school, to teach English. The package included fully furnished accommodation, paid airfares, and a large salary. I had tried to organise the interview on Skype, but it was just so strange because the interviewer refused to call on a video chat but rather opted to communicate by writing messages on Skype. The most bizarre and strangest thing was that he sent me the contract for the job and literally said, "Read it and sign it and you have the job!" There was no visual communication between us, so I was a little uncomfortable about this, so I just forgot about it.

The United Arab Emirates

Now this job seemed to be the most realistic and promising of them all. One day while working in the station, an ATM engineer was telling me about this company which originated from Lebanon. A school which had many locations across the Middle East and Arab world. The most appealing thing about this school is that they did not require applicants to have teaching experience or even a teaching related degree. All they required candidates to have, was a university degree in any discipline. You could have a degree in engineering, and you would be accepted.

Following an exceptionally long application process and Skype interview, I made it to the final application stage, the face-to-face interview which was held in a Dublin hotel. I went all out for this interview because it was the job that I wanted so badly. I even purchased a new suit specifically for this interview. Upon arrival for the interview in Dublin, there were so many people here for the same thing. After maybe half an hour waiting, I was called but it was so strange, the interviewer, a Lebanese lady was interviewing three of us at the same time. Not only did it feel bizarre, but it also felt so unprofessional, because every question she asked we all tried to outdo the other with our response. The best was from the girl who was being interviewed aside me when asked, why do you want to teach? She flashed her large eye lashes and replied, "Ever since I was a little girl, I knew it was my destiny to be a teacher, to educate and contribute to a child's learning, and that is what I was put on this earth to do," and she was making hand gestures and had a very false smile. I thought to myself, *What bullshit, how the hell can I match that?*

We then ended the interview and were told we would hear in a few weeks. But there were many applicants there that day, and it was the second day of it, they still had another day to get through. As a matter of fact, my mother's friend, her daughter had an interview the previous day. So, I was able to talk to her and get some advice on what to expect from it. This interview, for me, felt very intimidating and uncomfortable. There was just no possible way that they were getting accurate replies, with candidates constantly trying to outdo each other.

This interview was held around the middle of March, and then I got a reply maybe two weeks later. Sadly, I was unsuccessful in my application due to the fact there was an overwhelming number of applicants and there were insufficient positions available, but I was to be kept on file should something come up. I was told if you made it to the physical interview, you were 'hired', and they were going to place you then in a country or school. But you know, I feel it was the best thing that ever happened. I did not realise the world of opportunity that would waiting for me in China.

Then, there was China

I had to register on a TEFL job website to apply for the positions previously mentioned. It also required me to fill in a detailed profile which would be visible to schools and organisations which look for the right applicants. While applying for the Emirates position, one night in November 2015, I received an email from a school in the city of Guiyang, the province of Guizhou in the Southwest of China. The email stated that I was chosen as a possible successful candidate to take up a position in a private school to teach English. I exchanged information with Chris, who arranged a Skype interview the very next day. He was very keen to talk to me. We conversed for perhaps thirty minutes. He was telling me about the school and the city in which he spent many years in. He was interested in the detailed course I was studying online. He was hoping to get me out to China in February but, this could not happen because I was not prepared to go so quickly. Too much was happening, between studying to finish my course and attending a psychotherapist to get help with dealing with my anxiety and stress, and to wrap things up in Ireland.

It was such a relaxed interview it almost seemed too good to be true. We decided to end the conversation with me finishing my studies and getting back to him as soon as I finished the course in March.

Now I must tell you, I had zero interest in this position. I said I would have the interview just for the experience of getting used to interviews for teaching positions. The thought of me in China made me laugh and seemed so unrealistic. It was not just me who found the idea of me teaching in China humorous, but some of my friends and my family too. I felt that this would not be me and that there was no way I would consider it. I decided to scrap the idea of going to China and focus on the Middle East.

It was one evening before the email came through with the news that I did not get the Emirates position, that I started thinking. '*What if I was to use this job in China as a steppingstone to get the experience for the Middle East?'* I was a little more interested in what I called 'Plan B'.

All I need to do is get one year's experience in China, then apply through the same company as I did before, except this time, my application would be strengthened by relevant work experience. Strangely enough, I received another email from Chris literally the same week as the email came in about my unsuccessful UAE application. Indeed, did he stay true to his word. He asked me if I was still interested in the job and if I was finished my TEFL training. I was still interested in this job, and I was just finishing my TEFL course, finally. So, we arranged another quick interview that week to just review the position and he asked me to send him some documents which were required.

The very next day, I received an email from him with 'job offer' as the subject. I had been successful in my application for a job at his school. I was in my office, and I had just got back from a food safety training course and realised that shit had just got real. I was overwhelmed and over the moon, for here I was just after receiving the news that made my dream become reality. Even though, it was not the original plan for what I wanted. I didn't mind too much because even though I took a different route to my goal, the journey is always the best part, the process of growing, living, and learning is much more remembered than the destination.

Now, I had something else to do, to break the news to my parents. I knew this was going to be such a bitter and hard pill for them to swallow, because it would mean their youngest son going off again across the world chasing his dreams and searching for a new life. Anyone who is a parent does not really want to see their children leave home.

They had so many questions and concerns about the job and the place I was going to, typical parent concerns. When they asked me where exactly I would be going to, I said, "From what I know it is in Southwest China, in a city called Guiyang which is like left and west of Shanghai." So, this was my answer to which my parents replied with a blank face asking me "So, you are going to a city you never heard of, and you don't know where it is?" That was the conversation I had with my parents!

In addition to that, I could not even pronounce the name of the city that I was going to!

It took my parents a day or two to come to terms with the fact that I would be flying the nest again, but I will point out they are the most supportive people I know, without them I would not be the person I am today. I credit them for everything I am and everything I am doing.

I kept the news of my new and exciting life under strict wraps for about a month because I had not told my work about it, and I didn't want to make it public knowledge just yet. I needed time to get organised and to get my affairs sorted. When the time came to break the news to my new area manager, I arranged a meeting with him. I feel I played this very well and it would be something worth noting for yourself should you ever consider doing what I am doing.

I asked for a year's leave of absence, a career break to pursue a dream. I told him that I had studied to be a teacher almost every night for the previous eight months. He was surprised by this. I have to say, I could not have had a more supportive area manager in this situation. Had it been the manager who caused me so much trauma during the previous year, it would be a given fact, he would not entertain the idea of me doing this, he would refuse my request.

Within the next few days, I received a phone call from him, in which he had discussed this with the HR department of the company, telling me that he could only get a six month leave of absence for me, but to keep it quiet that he would get it extended after the six months for a further period. During my time there, I had built up a strong relationship with the HR department and the department head who, I was told, did not hesitate in granting me the leave.

Now, to explain the reason why I asked for a leave of absence. I decided I wanted to keep the door open on my position, and not close it out should something go wrong, and my teaching opportunity fell flat. Having something lined up as a 'just in case', can be so good to fall back on if something happened. I knew in my heart and soul that I would not fail at this new venture, and that there was no way on earth I would ever work in retail again, but at times there is no problem in keeping the door open, in case of something going wrong.

I was able to extend my leave of absence for the further period in late 2016, which meant I would return in the summer of 2017 to my old job and old life. The life I so desperately wanted to leave behind, inevitably, back to my stressful job. On Saint Patrick's Day 2017, I decided to break the chain that connected me with my old life. I sent my resignation to my area manager stating that life has taken an unexpected upward turn for me and I was not ready to leave and that I decided to spend more time in China doing what I felt what I was meant to do. I got a positive reply wishing me all the best and to contact him if anytime I needed anything. I appreciated that.

It is now 2022 and I am in my sixth year living in China. Being here has provided me with so much growth, knowledge, and experience. It seemed at the beginning of my TEFL career it was the Middle East and nowhere else, how times have changed. I no longer desire to go there for work, I am quite happy and content staying in Guiyang. China has become home to me now. The people, the places, and the food, it is too much for me to walk away from now. It truly opened a world of possibilities and opportunities for me. It is crazy how things happen. I changed the plan, but I still hit my goal of living abroad. If you are enjoying something, well keep doing it, do not fix what is not broken. I have been truly blessed with friends that have become family to me, and so many opportunities that I could have never thought of having at home in Ireland.

So, if this chapter has taught you anything, I hope it will be these two things.

Number one:

No matter what your goal is and no matter what your plan is, there will be times in which difficulties and problems will force you to change your direction. A goal that is easily attainable, I feel, is not a challenge. I like a challenge and China has provided me with such great challenges that I have enjoyed. My goal at the time was to use China as a steppingstone, to get to the Middle East. But you can see how it changed.

Number two:

As billionaire and philanthropist Warren Buffet says,

"Never test the depth of the river with both feet." This quote sums up how I feel about taking risks. I learned always have something to go back to if something you are doing falls flat. This will allow you to pick yourself up and give you the chance and the opportunity to start climbing

again. I feel it is a good idea, because you may never know down the line when you might need to contact previous employees or businesses.

Perhaps you are setting up your own business and need advice or even a place to get your first taste of success, it is good to start with where you once did and go from there.

That's why it is so important to develop good, strong, respectful, and trustworthy relationships not just with your family and friends, but also with your employees and employers. You never know when the time comes that you need to reconnect with them, and if you leave on a good note, they will do what they can to help you.

Chapter Seven

Isolation

I arrived in China on the 22nd of July 2016, landing first into Shanghai, where I had to wait for a few hours for my connecting flight to Guiyang. While here, I had my first culture shock experience. While in Pu Dong International Airport, I looked around and I could not see any other foreigners around me. Just Chinese people. I felt excited because I was out of my comfort zone.

After a delayed flight, I finally arrived in Guiyang at around 1am.

I started my first week in China with a bang. It was indeed a rather dramatic and worrying one. I had begun to settle into the routine of the school and the way of life. When people arrive for work in China, they must undergo a medical examination if they do not pass this, they cannot get a work visa to stay in China. My medical was done on the Tuesday, and I was all set to get stuck into work. I got a call from my Chinese boss on Thursday, informing me that I needed to go back and get another blood test at the hospital. I thought maybe they noticed something with my thyroid because it fluctuates up and down all the time. I went to my school and my boss drove me out to the hospital. I gave blood again which I do not really like doing, because I hate needles.

After I gave the blood, a doctor came and took me upstairs to an office. Here he looked at my medical report and then opened a translator on the computer. He asked the following questions:

"Is there a history of liver disease in your family?"

"No," I replied.

"Do you drink alcohol excessively?"

"No," I replied.

"Is there a history of hepatitis in your family?"

I hesitated in my reply to this question because I immediately thought of my dear grand-aunt Marie, who had suffered with hepatitis, which developed into liver cancer and inevitably killed her.

"No," I replied.

"You have hepatitis B."

You do not need me to tell you how this made me feel, I freaked out. I immediately went for my boss who was sitting downstairs. I brought her up, I had tears in my eyes as I told her about this news. She tried to calm me down, but she was wasting her time. She then spoke to the doctor who told her to get me to another hospital to get it checked out immediately. We then left the hospital and returned to the city centre.

On the way, she tried to reassure me that everything would be okay. I told her that this was a mistake and that there was no way I could have this disease. She asked me I should not drink so much alcohol, most foreigners here like to drink. She had a shocked look on her face when I told her that I did not drink any alcohol. She kept saying everything would be fine, and that I had nothing to worry about. I looked at the date, it was July 29.

Of all the days to get news like this it had to be this day.

July 29 was the anniversary of my grand-aunt Marie's death who died on this date in 2013.

On the way in the car, I searched the symptoms and causes of hepatitis, to check if I had experienced any of them.

Some of the most common causes were:

1. Excessive drinking of alcohol
2. Excessively taking pain killers
3. Unprotected sexual contact with the same sex
4. Sharing dirty needles

None of these were relevant to me. I was still wondering how could I have this? Me? I live a healthy life. I train, I eat clean, and I respect my body. Perhaps, I drank from a cup or used something that was infected by someone who had it. I did not know.

We arrived back in the city, and she took me straight to the Guizhou Provincial People's Hospital for more blood tests. We had to wait over an hour to get into the liver disease unit due to the large number of people waiting for an appointment.

After waiting, we got to see the doctor where she told the doctor about my condition and what had just happened. He said not to worry. This was starting to irritate me. I began to worry, and I thought to myself, *'Of course I am going to fucking worry. I could have a disease which is going to have a huge impact on my life and everything that I would do in it.'*

After they spoke, she told me that this was the best liver doctor in the city and that he would take care of me. He told me to come back in the morning for more blood tests. After which we left and then I met Chris.

My boss had contacted him and told him how I was feeling after what had happened. He took me for coffee near the school in which I immediately told him that there was no way I would stay in this country if I had hepatitis disease. I said I would have to go home and get treated there for it.

As someone who suffers with anxiety and stress, it is not easy to bring yourself back so fast. I could not stop worrying about it. My next plight, how would I tell my parents? It came quite quickly to me; I would not tell them. As soon as I told them, they would insist I get back home immediately.

I decided not to breathe a word of it until I had a clearer picture of the whole situation. When I got back to the school, I spoke to Chris, and we agreed that this would remain strictly between us, and we would not tell the other teachers about it. But just because we would not speak of it did not mean I escaped from experiencing the isolation.

My boss had the responsibility of taking some measures to keep me safe and to keep the others safe. To prevent the spread of the disease and contamination she segregated me from the others. Now, when I mean segregated, I mean I had to keep my bowl, glass, chopsticks, and plates away from everyone else's. They were kept in the kitchen so that no one else would take them or use them.

Later that day, we were having dinner and after I finished eating, one of the other teachers and one of my very close friends, Paul, went to take my bowl. I refused to give it to him and said that I would bring it out myself. He looked at me funny and let me carry on. He asked me later was I okay, I told him that I had a bug and that I did not want anyone to catch it especially being around kids so much. He understood, which was a relief for me. The last thing I wanted was to be treated like a leper by my new work colleagues. But I started to feel like one. Every few days,

I had to return to the hospital for results and to talk to the doctor. Very slowly the results were coming back as negative which was relieving the stress from me. One night I went for pizza with Paul, and I told him my current situation and I apologised for lying to him, but he understood why I did it. He was very concerned and was very supportive and sympathetic towards me.

He asked me if I had told my parents I told him no I did not. He said that it was the right thing to do. I was so happy that I could confide in someone, at least I did not feel too awkward, someone knew and helped me kept me occupied while I was waiting for the test results.

This plight lasted three weeks. There was one final blood test result that I needed to get in which I had to go see the liver consultant with my boss. It was a Friday and there were so many people there waiting to see surgeons and doctors at the hospital, that we had to get our appointment ticket and leave for a few hours and come back at lunch time because there were over seventy people waiting ahead of me. Prior to going to the hospital, I took my rosary beads with me in my pocket.

Now, I am not a religious person. I have distanced myself from the Catholic Church, but if there was ever a time that I needed some sign that I was going to be okay it was then. I also had a picture of Marie in my wallet, and I prayed to her before I went to the hospital, asking her to make sure everything would be okay and that this was all just a huge mess-up, but the word 'fuck up' is more appropriate for this.

We went into the doctor's consultation room where we met the specialist, and he sat me down. He looked through the computer and found my results. It was clear. I did not have Hepatitis after all. My boss took great delight in telling me the marvellous news and she was so happy. It was such a relief that after the three weeks of stress and anxiety, feeling like a leper, I had nothing to worry about.

I just couldn't believe that such a fuck up could be happening. Perhaps the bloods did get messed up (that's what I think happened) and now there could be some poor individual who has this horrible disease and they do not even know about it.

Now, should this be occurring in Ireland, England, or America, without a doubt, there would have been a huge case against the hospital. But this is not Ireland, England, or America, it is China. So, things are dealt and handled much differently. I did not care about filing a complaint against the hospital or the doctors, I was just so relieved that I tested negative for hepatitis.

Words could not describe the feeling on that Friday when I got cleared. My boss said to me that I need to take great care of myself and eat more and exercise more.

I now found myself with an even greater task, telling my parents. I felt bad withholding this from them but really, I had to. It was bad enough trying to manage it on my own, in a foreign country and with no one of my own around me, the last thing I needed was to have my parents demand I return home. I was not selfish in my decision, but I learned that you cannot always tell them everything immediately, when you do not know the full story, and besides, it ended up being nothing.

So, I picked up the phone one night after and asked my mam and dad to sit down. Then I told them that I had something important to tell them. I just told them, I didn't tell them the whole fiasco of what I went through, but I did tell them before I started, that they would understand why I didn't tell them this and why I kept it to myself. They did not think any less of me for doing what I did.

I told them there was nothing to worry about and that if there was something seriously wrong, I would be back home as soon I could get a flight. After I told them I felt the weight of the last few weeks disappear so fast. There was nothing more to worry about. The sleepless nights, the worrying, the need for reassurance had quickly diminished from my mind. I was no longer a leper, nor was I in isolation anymore in my school. My problems were over, and I was getting ready for this exciting new chapter, this new road I was about to embark on.

But something heart-breaking was about to happen to me during my first week of teaching. Had I not suffered enough during my first month here in China? On Saturday August 27, 2016, again, a feeling of isolation.

Chapter Eight

Out of My Control

"For Val, who was always there,

and

who told me who nailed Jesus to the cross."

From my experience, perhaps the worst thing that can happen when you live abroad is when something happens at home to your family or friends. You are not there. You cannot help. It is not as easy as simply jumping into the car and driving ten minutes like I used to do. Now, I am on the other side of the world, and when tragedy and loss hits, you cannot help but feel alone and helpless. It is simply out of your control.

Before I embarked on my adventure to China, I had a list of things that I needed to do at home. On this list was something that I added quite late, but it took priority immediately. Just a few short weeks before I left Ireland, I planned a weekend trip to Blackpool with my brother to see our cousin Val and his wife Cathy. Time was really against me in the run up to leaving home, I was still working, but this weekend with Val and Cathy took priority over everything else. Val was Dad's first cousin on his father's side, but they were more like brothers. Dad's brothers were considerably older than him, a huge age gap which meant by the time Dad was growing up, they had already flown the nest and went to

England. Val was just one year older than my dad and they grew up together.

Val was more of an uncle to us than a second cousin and we loved him very much. Earlier in 2016, Val was diagnosed with cancer, and he was receiving treatment.

To be perfectly honest, cancer is a bastard. It does not care if you are black or white, gay, or straight, or rich or poor. It will take you and kill you. I only know a handful of people who are fortunate enough to beat this evil disease. I am not trying to sound negative about this, but personally, with the amount of people I have lost to cancer over the recent years, I have lost faith and belief in there ever being a cure or hope for people. Most of the people I have lost, family members and friends fell victim to the disgusting disease.

Val was always there. Birthdays, communions and confirmations, and there would always be a card. We would always look forward to his annual visit home to Callan, where often he would stay with us or stay with his older sister. But, regardless of where he ended up staying, most of Val's time was spent with us but he also spent a little time catching up with some people from his youth.

I have so many wonderful memories of him. Memories that still make me laugh with joy and delight that they happened. When I think of these memories, I do not think of losing Val, I think of the fun times, the happy times, which got me through the loss.

Val was a man of good taste, and he also had many talents. Cooking, photography, and a strong sense of fashion, to name a few. I recall one memory collecting him in late 2015. Val had arrived and he went into Callan to catch up with some old friends in one of the bars. So, I was there waiting for him to come out of the bar, I had not seen him yet on this trip because I was working.

Now, I knew straight away what condition he was going to be in! I parked outside the bar, and I waited a few minutes and then I saw a distinguished looking man with a black Adidas tracksuit, golden rings on his fingers, golden chains on his neck, and with a black baseball cap on his head.

He jumped into the car, and we went out home. As we entered the house, he brought with him a shopping bag from the local supermarket and then he walked into the kitchen and pulled a huge leg of lamb out of it and threw it on the table. He said to my mother, "Don't worry Helen I brought dinner with me!" It was his attire that took the biscuit, only Valentine Francis O'Brien could of pulling that attire off!

So now you get a taste of Val's wonderful and fun humour. He really was a gentleman.

So, after our flight arrived at Manchester Airport, we took a train which brought us to Blackpool train station and there Cathy was waiting in her car for us. She was so happy to see us, and we were so happy to see her too. She told us that Val was so excited that we were coming over to see Cathy and him, that this would be the best weekend of the year for him. She also warned us that we might get a shock when we see him because he had lost so much weight with the cancer and treatment. Meeting cancer patients was something that I sadly grew accustomed to particularly in 2016, because I had lost another dear friend only months before.

A short ten-minute drive brought us to their home, and we walked in. There he was, the man we looked up to, the man who was always there for us, sitting in a chair restrained with tubes from his nostrils connecting to a drip. We went over and hugged him; he was smiling. Cathy said it was such a long time since he had smiled last.

So, we then chatted for a little bit and then we gave them gifts that we had bought for them at the airport. I had brought another gift for him. I brought a set of rosary beads made of olive wood that I had bought in Jerusalem just a few weeks prior. I had told him that I placed them on the stone where Jesus' body was washed and prepared for burial. When I handed him the beads, he clutched them in his hand and started smiling which then turned to tears.

I was so worried that I had upset him, but I did not. He reached out for us, and he hugged us tight. Our eyes were filled with tears too and so were Cathy's. She told us that he was so looking forward to seeing us that he was overly excited and very emotional that day.

We then laughed and chatted late into the night, and I think maybe near one o'clock, Cathy took us to our accommodation for the night. Val and Cathy had an incredibly beautiful mobile home, so we stayed in their other mobile home just a few minutes away from them.

We spent the next day together talking and laughing. Both Val and Cathy were so excited and interested in China and particularly, the food. They said they always wanted to go there. They were looking forward to following me on Facebook and seeing the photographs. We drove to a country bar outside of Blackpool, where we enjoyed some tea and again, we talked for almost two hours.

Conversations with the right people are priceless.

After we returned to their home, we had dinner, and some of their friends called over and it was so wonderful to meet them. We also met

Cathy's sister and her family. Again, that evening we spent at Val and Cathy's home talking and laughing and learning about our family. Val had a vast knowledge of who was who in our family and he was great to go back and trace relatives. I learned so much from him.

Val and Cathy also informed us that they were planning to come home to Callan for the last weekend in August. There had been a World War One monument, which had been constructed in memory of the men and women from Callan and surrounding areas who served in the war. My great-grandfather and my grand uncle in-law served and fought in that war. So, it was very meaningful for Val to be there, and to be with Dad to honour their names. The unveiling was to be held on Sunday, August 27. Val and Cathy had made it their goal to come home for this event.

Monday afternoon came, and we were preparing to leave them, but not before we got a glimpse of Val's extravagant hat and boot collection. Val had a huge collection of hats, various designs and styles from all over the world. His latest hat to his collection was a Cuban style hat. He recently had it delivered and was so happy showing it to us. We also saw all his cowboy boots which he had bought on his many visits to America. He gave us a pair of boots each, along with a beautiful photograph of the bridge in our town. A photograph that he took on his last visit. It was a large photograph, so we had to carry it as carry-on luggage, and it had to be handled with great care.

Val and Cathy drove us to Manchester Airport which was situated not far from their home. I knew what was coming next... The dreaded goodbyes.

I said to myself I was not going to cry when I said my goodbyes to Val. It was so hard to hold back the tears as I put my hands around him and hugged him tight and told him I would see him and Cathy in China very soon.

This was something I knew well would not happen. But that is the thing, we always keep faith and stay positive and hope for a miracle to happen. I knew deep down inside of me that this would be the last time I would see him. Val also had tears in his eyes as he hugged me and my brother. After a few minutes, we finished our goodbyes to Cathy and Val and we both turned and walked into the airport. I saw them drive off and they both waved at us, and I said to myself, *'That will be the last time I see that man.'*

I apologise to you for having this attitude towards cancer. It is just that I have seen so much of it (we all have), and it is so difficult to remain positive while the disease kills our loved ones.

My remaining days and weeks passed quite swiftly in Ireland as I prepared for my move to China. During this time, I finished my job as manager at the gas station (a leave of absence). I had my going away drinks the night I finished the job. I spent the last days with my family and visiting some friends.

The night before I left, I made calls to my aunts and uncles in London, and of course dear Val and Cathy, who wanted to wish me all the luck in the world. They said that they were excited and looking forward to all the photos and the stories.

I landed in China and got settled into my new environment and after a few weeks spent dealing with my blood problems, I started teaching on August 24th.

Friday, August 26

I had just finished my third day at school, and I was loving it so much that I sent a message to Val and Cathy telling them about it. They were thrilled to hear the news. They also told me that they both were preparing to come home to Ireland that following day for the unveiling of the World War 1 monument in Callan.

Saturday August 27

I woke early to meet my colleagues for coffee and to go to school. As I sat in my apartment waiting to leave, I scrolled through my newsfeed on Facebook, and I saw that Cathy had posted 'anxious' as her status on Facebook status. I immediately contacted her and asked her what had happened. She informed me that Val had fallen and cut his head. We left it there because I knew she had to take care of him, and it was also midnight in England (there is a seven-hour difference in the time zone).

I had completed my first class from 10:30 am to 12:30 pm and was having lunch with my colleagues. I then started my next class at 1 pm. At around 2:50 pm, I took out my phone to see the time and noticed that there were several missed calls on Facebook and on WhatsApp from my brother and my mother. I started to get concerned and then looked at a message from my brother and it said, "Ring mam as soon as you can." By the time I finished the class, I had received a message from Cathy on Facebook that Val had died.

Val had a heart attack at home and the paramedics did what they could for him as they took him in the ambulance to the hospital. He was gone.

I immediately went to the teacher's lounge and broke down crying. The other teachers came to my side, and I told them that Val was gone. My colleague Chris, said to another teacher, "You need to take Philip's next class." I cried and wailed until my eyes were red and I started to cough. I went into another classroom and just let it out. I then took my boss's phone, which had access to international calls, and I rang my parents. Dad was not able to talk as he was crying. I spoke to my mother who was also crying and asked me how I found out. I told her that I was in class and got the messages and then Cathy's message came through. I spoke to my father briefly but with us both crying it was hard to understand each other. After a few minutes, I told them I would call them later. I returned, red-eyed to the lounge where one of my colleagues comforted me and told me to cry it out.

Even though I knew that this day would come, that we would get the call, we were never prepared for that day. We hoped that it would never come, but it did. My heart was broken. I experienced heartbreak only a couple of times and this was one of those times.

What made the situation worse for me, was that here I was, all alone at the other side of the world, while my family were in Ireland and in England. Who did I have here to turn to for consolation? Nobody. Just my colleagues and friends that I made over the last month from living in China. One of my dearest and closet friends, Courtney, did not know what had happened until after about an hour or so after I got the news, and she came over to me and gave me a hug. She also had tears in her eyes as she hugged me. My colleagues knew how much this man had meant to me. They were terrific in their support during my grief.

My boss was also incredibly supportive, she told me that my classes would be taken from me for the Sunday and that I did not have to worry about teaching, just take time and take care of myself. I really appreciated the love and support that I received that day. It is something that I will never forget.

I went back to my house that night and contacted my parents again. We were still crying when we were talking. They were feeling terrible as they realised that I was over here with nobody to turn to, only my colleagues. It is times like this that I wished I were back at home.

I made the decision that I wanted to return home for Val's funeral. I transferred the money for the flight to my bank. I was determined to return for it, even my boss that day had made calls to get my passport back to me, as it was sent off to get my residency permit, so that I would be able to legally work and live in China. The benefits of having a boss with contacts. I stayed on my own for the remainder of the evening shedding tears over Val. Then, I looked at photos of us together that I had posted on my living room wall and after a few minutes I started to smile, and I laughed as I remembered the good times we spent together.

From remembering the times, I have spent listening to his stories, to going off through the fields with Val, Dad, and my brother, and of course him telling me who nailed Jesus to the cross. These were the memories that I had of Val, and they were all I had to comfort me as I sat there, alone.

The days that followed...

I woke the next morning, staring at my ceiling, thinking about the trauma and the heartbreak of the previous day. Was it all just a bad dream?

Unfortunately, it was not. Reality had hit me in the face, and I looked at my phone, at the messages of condolences I received from friends and family. I went to my school for around 10 am, even though my classes were taken off me for the day. I felt I wanted to be around people and to keep myself busy, focused on something else and not on the sadness.

I got to the school, and everyone asked me if I was okay and if there was anything that I needed. I really appreciated that. My boss informed me that they would have the passport for me on Wednesday morning. I had to return to my home to get my passport receipt (this was required to retrieve my passport from the police station).

As I waited for the bus to return home, I started to feel sad, and as I got on the bus with my head low and with tears in my eyes, I looked up and there was an elderly woman looking at me and she smiled at me. There was a vacant seat beside her, and she beckoned me to come over and sit down. I sat down next to her, and I smiled at her and said thank you.

The tears had dried up from my eyes and I thought to myself what a kind and wonderful gesture from a stranger. She looked into my eyes, and she just smiled. I was so grateful for this lady's kindness to me, a stranger, a foreigner. She knew there was something wrong and this simple act of kindness was her way of consoling me. I will never forget that lady and the kindness she showed me.

I spent that evening with my friends and colleagues. We went to a bar, and I was so grateful for them. They helped take my mind off everything. After a few hours, some of us went to a KTV (a karaoke bar which had individual booths where we could sing in private). I met some new people who had arrived to teach at other branches of our school. We stayed there until the early hours of the morning and around 3am we decided to return home.

I spent the Monday at home. I went to the gym in the morning, but it was extremely hard for me to get into it. Even though I am usually always focused, this time it was so difficult for me to do anything. For the next day, I had a trip with a girl that I had met in a bar the previous week. We arranged to visit an ancient town located outside the city.

She told me where to meet her and I went to the location. She met me outside the location, and she took me to her car. When I got into the car there were two men sitting in the front. I did not know what was happening and she told me that she was bringing her friends here too (it felt more like bodyguards rather than friends. I thought it was just going to be the both of us.

To be honest, I did not really want to go on this trip, but then I thought it might help get everything off my mind for a while. For most of the day, the girl and her friends did not really speak to me, the three of them were ahead of me walking and I was left walking behind them just looking around. The girl spoke some English, but the two guys did not know a word. It felt uncomfortable for me, and I started to wish I did not bother going on this trip and my mind started to go into thinking of Val and starting to feel down about him.

After walking around the town for most of the day, they decided to go up and see some wind turbines. I had no choice but to go with them. It was starting to get late now, almost 7pm, and I was asking them when we would be leaving, because I had to go to work the next day. The girl told me that they were considering camping here for the night. I told them that there was no way that I could do that and that I had to get up for work. The last thing I wanted to do after spending a miserable day with strangers was to stay away with them for the night.

We returned to Guiyang around 9:30 pm and it was raining. She asked me where I lived, and I told her. She told me that they would drop me there. A few minutes later, they stopped the car and told me to get out. I just looked at them and I got out and they drove off. I was nowhere near the complex in which I lived, and I was on the other side of the city. I stood in the spilling rain, but deep down I was happy to be away from them, so I grabbed a taxi, and I went back home. What an experience, seriously, it was the last thing that I wanted to do but thinking that it could help me to change my mood and keep my mind busy, soon turned out to be a bad idea. Lesson learned.

That Wednesday, when I got to school, my boss gave me back my passport and told me to let her know when the funeral arrangements are confirmed, and she would sort out cover for my classes. I was happy to be back at work, focused on something to do and to be with people.

On Friday, Cathy informed me of the funeral arrangements. Val had previously made it known of what he wanted. He wished to be cremated and his ashes brought back home to Callan and to be buried with his family. Cathy had set out to fulfil his final wishes. His ashes would return home later.

That night I started to search for flights, and I found a reasonable flight home. I was very sure that I wanted to come home. I spoke to my mother, and she told me that it would be such a long journey to make. She also told me that if I did come home, it could unsettle me, seeing my family grieving and so upset, it could make me want to come home.

I spent some time thinking about what my mother said, and I decided not to come home for the funeral. I realised that I already had said goodbye to him in July. I did manage to be part of the funeral, as Cathy asked me to pick a song that Val would have liked. I chose a song called *The Parting Glass*. It did not take me long to decide on it, as thing was a song that reminded me of him so much.

My father and brother went over for the funeral in the middle of September, it was on a Wednesday. What a horrible day I had at work. Val was cremated, holding in his hands the olive wood rosary beads that I had gotten for him earlier that year when I was in Jerusalem.

Val's ashes came home in November that year and were placed in the grave with his parents. Val was home and I knew that when I came home, I could go and visit him.

Losing someone close to you is very painful, it is twice as painful when you are on the other side of the world. It was one of the most painful experiences I ever had living away from home, but I must say, it made me much stronger. It made me look at things so much differently, that nothing is in our control. We must accept everything and make the most of our time and spend it with those we love.

I will end this chapter by giving you some advice. If there is someone you love dearly, and you do not spend much time with them. Go and spend time with them because you never know when it will be the last time. Take my advice because I had the chance to spend time with not just Val, but other close family members and friends who have died. You will regret not doing this.

Some people do not value this time and they take it for granted that people will be here forever. So please, do yourself a favour. Call that relative or friend and make the time to spend with them, whether it is taking them for dinner or sitting with them and listening. Because believe me, once they are gone, they are gone and there is nothing that you can do about it.

With Val in Blackpool. The last time I saw him.

Chapter Nine

Out the Window

No matter what we study for, or whatever length of time it is for, we all learn and gain theoretical skills that we plan to use in practicality. When some people get a new job, they sometimes bring a notebook of some notes and explanations that they took down in class while studying, that they feel they go back to and refer to when they are unsure how to do certain things, use it as a guideline, a bible. It is great to have such a powerful resource in your hand, a reference to help you through some obstacles or generate new ideas from. When I was working in retail, I kept a notebook on how to do different things such as invoicing and stocktaking.

Something that I have learned from teaching in China, is that there are experiences and lessons that are not taught from a book, nor from a teacher. The teacher of this is called experience and the classroom is called real-life. The materials are what you have around you to use, and the student of course, is you. I can honestly say that I have developed further in my teaching career not by applying what I learned from the 300-hour course I studied, my terrible knowledge of grammar rules, or my lesson plans which used to pinpoint each individual activity step-by-step. Rather that, I learned from the method of walking into the classroom and take what you learned in theory and throw it out the window, forgetting the majority of what you learned from the course and the books. The children who you teach do not care that you have a time frame for each game or activity, nor do they consider that you spent so much time researching the best way for them to learn how to say, "I am fine" or "I like apples".

Experience in the classroom has taught me that you must teach and communicate in a way that the students will learn effectively. Every age group or level is different and unique. It requires you to have a variety of means and ideas to be able to teach them in a way that they can learn.

Teaching English also requires you to be organised and prepared to a certain degree but there are times when what you want to do and what you want to teach does not go to plan. So, what do you do? Tell the class of twelve kindergarteners to "be quiet" while you consult your bible of notes form your studies to see what to do when what you are doing is not working? Absolutely not. You must adapt to the environment and do what interests them.

One of my biggest struggles in the first three or four months as a teacher was teaching teenagers. At this level, you are not using flashcards, nor are you playing musical chairs. I was struggling to engage them in any form of conversation or discussion.

I would dread my Sunday afternoon class because I would feel bad going into it because most of them would always be doing their homework or talking about their plight of being in a boarding school and having no time to do anything else but study. What didn't help was that during my induction training, I was told this level is the hardest to teach because 'they don't care that they are miserable, all they do is sit there with a frown on their face because they do not want to be there, and that you probably won't get much out of them but try anyway and see what happens'.

As you can see, I was given an extremely negative impression of teaching this age group.

We would attempt to read a book that I felt was not suited for this class. So, about eight weeks into teaching them, I slowly began understanding them, but they were not giving me much to go on. The occasional remark or answer to an open-ended question would make the class bearable for me.

I was getting frustrated at this, week after week. So, one day, while reading the book, I stood up and threw my book on the table at the far side of the classroom, told the students to do the same and to sit back down, then I took out a deck of UNO cards and spent an hour playing with them. I also took a bag of fruit and pringles out of my cabinet. This was perhaps the most important thing I did with these kids. I spent time getting to know them as individuals and as people.

You see, when you are trained for a teaching job, you are trained how to teach them by following a system and guidelines. You are not trained how to get them on your side and how to get to know the students on a more personal level. This is the most important thing that you need to know about this level in my opinion. Experience with this level of students, has shown me that they do not need a foreign teacher who will add to the stress of their studies, but rather they need a teacher who can be a friend to them for two hours a week and teach them about life, information and lessons that books sometimes cannot teach.

Again, going back to the title of the chapter, you need to take your training and your books and throw them out the window and do what they want to do. Come across as a friend, someone who can teach them about life, share experiences, interesting and funny stories, and events. Show that you care, make them forget about the pressures of their lives for just two hours a week. Once you do this, you will see and reap the rewards that make you so proud doing what you do. You find yourself discussing so much with them, having in-depth conversations, and discussing some controversial topics with them.

They trust you and confide in you some problems and obstacles that they are fighting with in their young lives. As mentioned, a lot of these students at this age live in boarding schools. They have little time to spend with their family or friends, no time to practice their sports or indulge in the hobby that they would love to do if they had the time. They treasure and appreciate the time that you spend in the classroom with them. So, it is important to make the class memorable and special for them.

My classes with this level have taken me out of the school, to ice cream parlours, escape rooms and to arcades. Sometimes, you need to get out of the prison that is a four walled room and go out into the real classroom, which is the real world, where everything is seen in practicality not theory and use that as your place of learning and teaching.

Time had passed in my teaching career, and I saw the social improvements and their personalities grow, and what was once my least favourite class, had become my most favourite one. I even made it clear that I would not extend my contract if this class were taken from me. I was assured nothing would happen to this class and I would remain their teacher. Even now, I am teaching this age group in high school, it is so rewarding. I believe that this is the age group I was meant to teach. The kids are amazing, their English levels are very high, and they are quite humorous. Just like the students that I am teaching at high school right now. I feel it is the most suitable age of students for me to teach and communicate with.

Another struggle, I want to talk about earlier on in my teaching career, is about the very first class I taught in China. It was the first day of semester and it was going to be a new kindergarten class. None of them had signed up prior to this, so this was merely a demonstration class of the product we were providing as an English school. No pressure at all for a brand-new inexperienced teacher who just arrived fresh off the plane, with knowledge only of what I had learned from my studies.

Right before the class my boss had told me that it was very important for me to deliver a good class to give a great first impression because they had not signed up yet and she wanted as many students to sign up as possible. To add more stress to my first-time teaching, the Chinese assistants arranged that the parents could also join the class and watch for themselves. Talk about no pressure, eh?

So, in preparation of the class, I made my lesson plan which outlined the list of activities and the material of which\ I was going to teach. I will never forget it, the theme for the class was giving the kids names and teaching them about fruits.

I had purchased fruits from the fruit stand outside the school to use in class. I feel it is important to use realia when teaching young kids, as often as possible because from my own learning experiences, I always learned better when I could examine and see with my own eyes, visual learning, as it sticks in your mind much quicker when you are exposed and in contact with it.

So, I entered the class with my plate of fruit and list of activities that I had planned to do, all set, eh? All possible anticipated problems were considered. Except for one thing, the duration of the class was an hour and a half, and I had gone through everything in my plan in about forty minutes! I used my initiative to think of activities that made the rest of the time go right and active. I would refer to the list of game ideas I made prior to the class to see what to do next, but as mentioned I ran out of ideas, so I threw it away and just went with it, doing what I could.

As it turned out, I had about eight or nine of the students sign up for my class, much to the delight of my boss. After the class, I looked at the grey T-shirt I was wearing, it was drenched in sweat from all the going and the activeness of the kids in the class. An equivalent to a workout in the gym.

You can prepare for most issues that can arise in the classroom, but you need to be prepared to stray from the plan and just go with what works for you and what encourages the students to learn. If you need to play ring a-ring-a-Rosie for twenty minutes and they are learning from it, then play it!

You might be dizzy, and the kids might be dizzy, but remember you are providing the students with a method, a path in which they learn, you are providing them with the steps to discovery. When the students take those steps and walk them successfully, that is when you know and realise what you are doing is making a difference and that you are getting through to them.

Chapter Ten

The Teacher Is Also the Student

I really love teaching my students, especially teenagers. It is rewarding for me in so many ways. There are times when we switch roles and they become teachers and I become the student. There is so much that they have taught me. In addition to this, some of my friends here have become my teachers too.

Life of a Teenager

"What does it mean to be a teenager in China?" This was a question that I asked one day while in class with my students. They looked at each other and one of them said, "Being a teenager in China is very stressful, even being a kid is very stressful." I just looked around at everyone else and they looked at me and then they looked down at the table. I asked her what she meant by that. The reply that was given to me shocked me; I could not believe what I heard.

The student informed me that being a child or a teenager in China, their focus and attention is on education, and that there is often limited time to enjoy other interests and hobbies.

If you want to get good grades and get to a good school, you must do everything in your power to ensure this happens. Some students spend their weekends going from class to class, learning more and improving their skills. I guess this is where we come in as an English school, providing the students with extra tuition to improve their communication skills, speaking, listening, reading, and writing.

Most students spend their weekends travelling around the city of Guiyang, going to different subject classes. Some have math classes in the mornings from 9 am, then to a music class for 11 am or 12 pm and then to English class in the afternoon. But it does not stop there.

For exam year students, some students must go to a 'cram school' where they take different subject classes which will help them do better in their exams and it can help them get to the high school or middle school that they are aiming for. They all aim for the best school in the city and province, which is Number One High School (where I currently work), but you can imagine the number of students who are in other schools and other neighbouring towns, cities and provinces who also want this.

Therefore, in my opinion this puts an unmerciful amount of pressure and stress on the kids to keep striving and keep improving themselves academically to meet the criteria they must hit. The students who spend time on the weekend going to other classes have little or no time whatsoever to enjoy for themselves or to spend with family or friends. Some kids do not even see their parents or siblings on their weekends, because of work and other commitments, perhaps a younger member of the family needs to go to another class located somewhere else, so the teenager or child must go to their classes by themselves.

I often saw many kids come to the school by parents or even grandparents. It's not that they do not take an interest in their child's education, it's just that they have other things to do, and this is the way life is here. Kids, I feel had become independent at a young age here.

Going deeper into the conversation I had that day; the girl went into more detail when I asked about the social skills teenagers have here. She told me that due to the intense educational system here, it can be difficult to make friends and to fit in at school. With everyone wanting to be the best, to be number one, causes resentment among people, classmates, and friends.

As a result, many students can be bullied because they are 'better' than others and because they get better grades.

As someone who was bullied for many years while in school (I was bullied due to my weight and hearing disability), I could relate to this. I could see that this class of students enjoyed each other's company so much and had fun together, that this two-hour class, once a week, was the only opportunity for the students to be themselves and for some, to have friends. This was the only opportunity for them to meet and to discuss some problems, play games and engage in conversation. I would see the smiles and the laughter of the students but when the bell rang, meaning the class was ending the smile would disappear and the laughter died.

For it meant, the time had ended for them to be themselves and to return to school for another week.

I felt so sorry for them, looking back on the education I received and the system we have in Ireland, I consider myself lucky that it was not so intense. Do not get me wrong, there was pressure to get good grades to get that place in university to study that course you wanted, but there was not the pressure of constantly having to be number one.

Reflecting when I was preparing to go to university, I was working part time in the local gas station on weeknights and weekdays. Some of my colleagues had taken a leave of absence maybe five months prior to the final exams to prepare for them.

When I was planning to do the same, my mam told me that both she and Dad did not agree with the idea of it. She told me that I needed to have something to get my mind off study even for a few hours a day, that immersing myself in study and not having any outlet or break would be very bad for my health and could jeopardise my plans for university. This reminds me of a story that my mother shared with me. I will share it with you too.

"Mam told me that she had a cousin who did nothing but study, even on Christmas day. He would take his dinner into his room and continue to study. He got the grades and the doctorate that he wanted but socially, he was dead. He could not communicate with anyone because he focused so much on wanting to be the best at what he did, that nothing else mattered, not even time with his family. The story did not end there for my mam's cousin. He went to Rome on a visit and when he came back, he was a different person. He shocked everyone by saying he was quitting his tenured job at university and going to Rome to become a Catholic priest. As my mam said, 'he snapped', had a breakdown, it finally hit him. Soon after he was ordained and living in Rome, the family got a call saying there was an accident at his living quarters. He was found dead in the bath. The cause of death was a gas leakage in a pipe in the bathroom, but his parents came to a different conclusion that he took his own life owing to the fact he did not enjoy it anymore giving the fact he was doing nothing but studying, he was not living."

This story often comes to my mind when I teach. I just hope that this is not the case for the students. The fact that they constantly study and spend time learning can have extremely negative implications on their emotional and mental health. It worries me quite a bit, but I guess this is a different culture to what I am used to, and I must accept and respect it.

One time, I returned home for a short visit to surprise my dad for his birthday, and I was asked to give a talk to a class in my old secondary school about living in China and my experiences. This was an extremely high honour for me. I was telling the students about the pressure Chinese kids have in school and I firmly told them to be grateful for the educational system we have in Ireland, that other places have it much more intense.

A Wasted Skill

When people learn a new skill or a language it is always presumed, they intend to use it for something, like another "notch on the belt" or another "string to the bow". This is not the case in China. When people here learn English here, they focus primarily on three skills, reading, writing, and listening. There are not many opportunities for speaking practice in school. From my experience from talking to Chinese people here, learning English is a "tick the box" exercise which basically means it is a wasted skill. They learn it solely for exam purposes, not for everyday use or communication. It is such a shame because English, well, any language can open a world of new and exciting opportunities around the world in many countries, not just in one country.

Soon after I came to China, I remember one day I was at the gym and I said, "Hello, how are you?" to one of the trainers in which he replied,

"Hello, I'm fine." This was the conversation between us for numerous times, until then one day I asked him another question in which he replied with the assistance of a translation application on his phone, *"I don't know, sorry, I cannot speak English, listen yes, write yes, read yes, but speak no."* He looked at me with disappointment.

This made me think that it was such a waste for people to learn English here and never be able to use it. Now after learning Chinese for some time, I can talk to him in his own language, and now he is learning to speak English. Four years later, we can communicate with each other.

Another time, I was in a parent teacher meeting, and I had one of my student's fathers in with me and my Chinese assistant who served as an interpreter for me and for him. He spoke to the Chinese assistant and told me that he can understand what I say but he cannot speak back to me, all he could do was listen.

Many people here are in this situation. It is a pity this is the case, but as I mentioned the students are just thought to understand it but never use it. It made me realise, in Ireland we are sort of the same with the Irish language. We learn it. We learn all four skills, but really it is not beneficial because we cannot use it for much outside Ireland, and to be honest, we also cannot use it for much in Ireland either. It is sad because this is part of our culture, our identity, and our heritage.

Some people need it, as it is a requirement for getting into certain Irish colleges and universities but apart from that it is a dead language. Only certain areas of Ireland are still actively using this language. So, in my opinion it is just the same in China except that the opportunities Chinese people have with being able to speak English are phenomenal, the world is theirs. Many of my Chinese friends here had the opportunity to work and study abroad in countries like Australia, New Zealand, America, Singapore, and England. They have mastered the skill of speaking English.

The English classes that the students have in their primary, middle, and high schools focus a lot on listening, writing and quite a bit on grammar. I never met people who knew so much about English grammar! They know much more than I do and probably more than I will ever know. From my experiences, here and from what I have seen, it is better for a Chinese English teacher to teach English grammar and rules. They do a better job than native teachers do. So, I will be honest and say that I try to avoid grammar teaching where I can! It is just I do not think it is drilled into us as much in our school as it is here for them.

My responsibilities as an English school teacher were primarily to develop their speaking skills, while also attending to the needs of their other skills (reading, writing, listening, dictation, and pronunciation). My students' parents would request that their children have more opportunity to speak, and to have more vocabulary. Now, some of the kid's speaking skills are amazing with even some sounding like natives and there are some that are really struggling with it, but this is what the job entails at an English language school, to help them improve.

Currently at the high school, I am teaching speaking English, IELTS, extensive reading, academic writing, and physical exercise classes.

What I have learnt from experience is that if there is a strong, friendly, positive, and trustworthy relationship between the students and the teacher, then the speaking and confidence develops so well. When you do not have this balance, if the student's perception of the teacher is distant and not friendly, then quite simply there is going to be a barrier and it will end with poor results. It is, therefore, crucial that we understand, respect, and appreciate each other's background and abilities.

Chapter Eleven

Lao Wai

I will be honest and tell you that I was a very ignorant person growing up. Ignorant in understanding what it was like for foreigners, the struggles, the sacrifices, and the path that they walked on. I guess it is also fair to state that I only ever saw one side of the coin when it came to foreigners.

Growing up in Ireland taught me that foreigners came to Ireland to take our jobs, houses, and money, that they should remain in their own country and be satisfied there. This was my tunnel vision and there was no other way, and it really did impact on my thinking and my perception of others.

But then there was the other side of the coin that I discovered, the struggles, the sacrifices, and the paths they walked on. How did I see the other side of the coin? Simple. I became a foreigner. It was not until I walked in their shoes, did I understand and appreciate their journey.

Australia

In March 2013, I left my homeland of Ireland and emigrated to Australia for work, for more opportunities and a better life. I had just finished university at the worst time in Ireland. The economic recession, which left thousands without jobs, their dreams crushed, and families torn apart by suicide and mass emigration to Canada, Australia, and New Zealand.

I broke my parents' hearts by leaving home, but I had to do what I thought was best for me. Perhaps the hardest thing I have ever done in my life to this point was not bidding goodbye to my parents but to my dear grand-aunt Marie. Marie was my mam's aunt and really, the sister Mam never had. I had been extremely close with her throughout the years. About six weeks before I left for Australia, Marie was diagnosed with liver cancer after suffering from Hepatitis B.

Saying goodbye to her as I left her house in Kilkenny broke my heart, especially when she told me, "Now go and don't turn around, keep going," as I walked out of the house with tears streaming down my face. I remembered what she told me. I got into the car and as Dad drove me out, I turned and there she was walking after the car with her hands out as much to say, "Please don't go." I have never experienced anything so difficult in my life as to say goodbye to someone who I may never see again. Even to this day when I recall those moments saying farewell to her, they bring tears to my eyes.

The moment that I arrived in Sydney I became a foreigner. I was no different to the Polish, the Estonian, the Romanian, nor the Latvian coming to Ireland in seek of a better and more fulfilling life.

When I started job hunting that very week, I prepared copies of my resumé, and the first place I went to was a European supermarket chain which has a huge presence in Ireland. I was confident that I would secure a position there because I had seven years of retail experience in a forecourt service station and a bookshop. So, I was well experienced for the job in my opinion. I also had the people skills and work ethic for it, so I basically had the job, didn't I? I felt very confident. So, what was said to me when I introduced myself to the manager of the supermarket?

He asked me, "Where are you from?" I said, "Ireland." He replied, "No. I won't hire any Irish people." Shocked, I replied "Why? I have all this experience and I am hardworking." He told me, "Because you are Irish, and you have a reputation of binge drinking and getting into trouble." I said that I was a teetotaler and that I would not ever be conducting myself in that manner. I was told, "I won't make any exceptions for you or your kind, because if I hire you, you will piss off in a month or two when something better comes along so again my answer is no." Speechless and shocked, I thanked him for his time, to which he replied, "I am sorry, but if you want someone to blame, blame your own people for ruining your opportunity." I walked out shocked.

I could not believe what he had just told me, he literally painted all Irish people with the same brush, and made no exceptions for the genuine, honest, and reliable ones.

Another experience of this treatment happened when I contacted a job agency. My aunt Molly had put me in touch with some old friends from South Africa who had emigrated to Sydney and opened a recruitment agency. I had what they might call 'a foot in the door', a contact, and a contact can do a lot for you because after all 'it's who you know'. Again, I was wrong. I had emailed them my resume and covering letter for their review and then I got the returned phone call with the feedback.

I was told that firstly, as a foreigner in Australia I appeared too educated and over-qualified for many positions (over-qualified because I had a university degree). His exact words, "Take my advice and remove your educational skills from your resume to improve your chances for any job, that way you will not intimidate other workers because they will perceive you as better than they are." Sounds like I had the luck of the Irish. A few nights after that I began thinking. Thinking of what an ignorant person I had been for the last eight years of my life. I would discriminate against others because of their nationality and the fact they were coming to my country looking for work. It was then it hit me, and hit me hard, I had become one of those foreigners doing the exact same thing.

I soon realised the struggle, the rejections, and the discrimination that these people were suffering in my country, and yet here I was, going through exactly what they did. I was not proud of myself after this realisation, it changed my attitude completely and my appreciation took a different turn. I saw the second side of the coin and it really did slap me in the face.

With my confidence dropping after numerous job rejections and subjection to racial comments, I found myself with very few options. I was now in my third month, and I had finally secured a job in a construction waste site, working on a conveyor belt separating raw materials from each other such as stone, wood, and glass. There was a huge problem here, there was a lot of asbestos passing in and out of the factory and I do not need to tell you the serious health problems asbestos can expose you to. I had to do something, it gave me a means to an end, to keep the money coming in. The money was fantastic, but nothing, no overtime, double time, or treble time could make my health a secondary priority.

I left the waste site the end of May, and I travelled west of Sydney to go to a vegetable farm to complete my ninety-day regional work which would secure my second-year visa to remain in Australia. For anyone considering working in Australia, this is a requirement to secure a second year working visa, but if you are lucky enough to get a good job, you can be sponsored, and you don't have to do this.

Now that was an experience. I spent three hours travelling west of New South Wales to the town of Canowindra, where I had secured a job.

Upon arrival to the farm on a Sunday night, I was greeted by the farmer who brought me to my living quarters on the farm, a rundown, cockroach infested mobile home. I was going to be sharing it with four other workers, two Italians and two Estonians. They barely spoke to me while I was there. I had to clean the house with the farmer before I even got the chance to unpack my stuff from the car.

I originally was informed that I would have my own room in the house, but that fell through and I had to share a room with one of the Italians.

I woke the next morning early to begin work. I was excited as this was something I enjoyed doing with my dad and uncle, planting and picking vegetables, I could not wait to do it. We were picked up by the farmer who drove us out to the field to begin work. I was handed a knife and was told to take off my gloves as I would not be able to grip and cut the lettuce correctly. Well within two minutes, I had cut my hand open and there was blood pouring out of my hand.

I told the farmer I had to get a bandage and he asked me a question – "Where are you from?" I replied, "The Republic of Ireland." He said, "You stupid Irish bastard, why the fuck did you cut your hand?" Stunned, I did not reply, nor was I given anything to wrap my hand with. I kept pressure until the blood stopped. One of the others gave me a tissue which didn't do much, but it was something.

I decided to ignore what he was shouting at me, but when his brother joined in, I could not hold my tongue. What they were saying to me was vicious and racist.

After we cut the lettuce, we were cutting broccoli and throwing it into the trailer while the farmer's brother was sitting in the trailer, watching us, and checking the lengths of the broccoli. He shouted, "Who cut this one? Was it you, Irish?"

I replied, "Yes, what's wrong with it?" He said it was too short. Then he shouted, "You good for nothing Irish bastard. You think you are all great coming over here, taking our money and our jobs. There are planes of you stupid Irish coming over to drink and take the bread out of our mouths." With anger I replied, "Well at least the Irish got our independence from England."

There was silence for a few minutes, and then he continued. I guessed there had to be something wrong with him, or he was just incredibly ignorant. I just couldn't understand how anyone could be so horrible, and the others just carried on as he passed racial slurs and remarks to them. I was told not to retaliate, it just makes matters worse, but there was no way I was about to stand there and be insulted by them.

Enough was enough, after four grueling days I packed my things into the car and drove back to Sydney at 3 am in the morning, I could not take it anymore. I arrived back at around 8 am, after making a few stops and almost nearly killing a few kangaroos. Kangaroos can be dangerous if you hit them because if they impale the windscreen, they can start kicking and I would not like to be at the receiving end of a kangaroo's kick!

A little while after I arrived back in Sydney, I began thinking of where I was and what I was doing here in Australia. I was beginning to watch my dreams and hopes disappear rapidly as I could not see anything positive here for me, no opportunity. I never felt so low in my life, my confidence was shattered.

To make life worse for me, the following week, my grand aunt decided to stop her chemotherapy treatment for her liver cancer. She had realised she had had enough of it and had no quality of life; it took so much out of her. After my mother broke the news to me, I went out for a drive for maybe two hours, most of which was spent sitting in the car thinking.

I returned home and went into my friend's bedroom for a chat and told him I had decided I was going to return to Ireland due to my grand-aunt's deteriorating health and not being able to see any positive outcome from living in Sydney. I had my return ticket, and I phoned my travel agent in Ireland, and he was able to get me back to Ireland a few short days after.

I decided that the time spent with my grand aunt in her final weeks was so much more important to me than sitting in my friend's house applying for jobs in Sydney. If I did not return, I would never forgive myself as this was time that I would never get back.

So, after three long months in Australia, I came back to recession struck Ireland, with little hope I would get a job so far. I was lucky because within a week, I had secured a job in a large grocery store chain in my city. It was funny that I emigrated to a booming country, a country of opportunity where I could not even get a job interview only to return home and secure a job within a couple of days of returning.

I also returned to Ireland with a different attitude, a more understanding and respecting attitude towards foreigners, as I had been in their shoes and lived the life they lived, the struggles, the depression of trying to find your place in a new country who look down on you and regard you as nothing.

First-hand experience is the best teacher you can have in life. It teaches so much more than any book or classroom can do.

China

Again, I found myself subjected to been called a foreigner every day when I made the move to Guiyang in the Guizhou Province of China to work as an English teacher in July 2016. Once again, I was subjected to it, but I was not met with the malicious and horrible remark rather I was met with curiosity, because many people in this part of China are not exposed to many foreigners. Because they often tend to stick in bigger cities such as Shanghai, Shenzhen, Chengdu, and Guangzhou, where the number of foreign expatriates is greater.

Being called a foreigner in China is different because the people do not insult you, nor do they appear to be racist (well not to my knowledge and I am here a long time now) towards you, more of a curious attitude. You will get stared at but nothing negative is intended by it.

I am so used to hearing *"Wai Guo Ren"* or *"Lao Wai"* (which is Chinese for foreigner) shouted at me, especially by little kids and some adults to which I reply, *"Zhong Guo Ren,"* (Chinese person) to which we both start laughing and continue. Nothing bad is said but rather turned to humour and shows friendliness towards each other. Then, they ask me to pose for a photo with them or even in some cases hand me a baby and take photos, they say thank you and continue as normal with their daily life.

To conclude my experiences as a foreigner, I am happy to say being a foreigner has taught me so many lessons in life, showed me so many things both positive and negative, but I am not put off by them. I started off in my youth with a bad attitude towards foreigners, but experience has shown me that the best times of my life were lived and experienced by being a foreigner (and still living and experiencing). Finally, I think there is nothing like being a foreigner. I have learned so many things, life skills that have shaped me into the person I am today. The best things and the best people I have met were experienced as a foreigner. I would not change a thing.

With my Grandaunt Marie, and Grandmother Josie before I left for
Australia

Chapter Twelve

Culture Shock

"The feeling of anxiety, loneliness, and confusion that people sometimes experience when they first arrive in another country."
(Collins dictionary definition of the term 'culture shock')

For anyone who has left their homeland in pursuit of adventure or a new life, they will undoubtedly be subjected to culture shock in some form or another. Culture shock may strike you in the form of food, customs, and mannerisms. There are many forms of culture shock. Throughout different adventures and trips in my life, culture shock has educated me and shown me many places through different perspectives.

Food

Previously before I moved to China, my family and I would enjoy eating Chinese take away on a weekly basis. The Chinese restaurants in Ireland have a wide selection of Cantonese, sweet, sour, mild spice, and black bean dishes to choose from. We had our favourites, mine always being Cantonese sweet and sour chicken with fried rice. I had been to many Chinese restaurants in Ireland, but not once did I eat food with chopsticks in any of the places I visited. So, as my adventure to China was rapidly approaching, I had a taste of the food, and I knew what to expect.

I was so unbelievably wrong about this on all levels! There is indeed no shortage of variety of food to choose from in China.

Every province in China offers different foods. Like Beijing is famous for of course, Beijing duck and in Sichuan, you can get many chicken dishes. Guizhou is famous for spicy cuisine. At first, it was hard to adjust to it but after a while, I did get used to it and now I really enjoy it. Perhaps, the spiciest dish I had was Chongqing hotpot. So much spice and flavour in it. One bit of advice to you the reader, bring toilet paper with you whenever you decide to eat this dish!

I remember my first day in China, sitting in the teacher's lounge with the other teachers eating lunch. I stared into my bowl at what I was trying to eat, and I remembered asking one of the teachers, "How have you survived here for six years eating this?" It was such a struggle in the beginning here to eat the food. Especially because most dishes include spice. Since I moved here, my diet has changed dramatically, and I can honestly say I have adjusted to the food here and I really enjoy it. The only thing that I cannot have much of is soy sauce due to my underactive thyroid. I have eaten some very bizarre things here in China such as pig brain, bugs, and century eggs (look them up!).

While I was in Cambodia in 2017, I ate scorpion, locust, snake, and tarantula. My family and friends back home were horrified and disgusted that I would eat these kinds of things. Simply because everyone knows how much of a fussy eater I am and that there are so many things I would not eat.

The one thing that I have gained attention for is my hesitation to eat mushrooms. It is the one food I cannot stomach. My 'Asian food' tour gained some attention, and I was subjected to fun attacks from people saying that I would eat a scorpion or a tarantula but not a mushroom!

Briefly, I will give you the reason why I hate mushrooms so much. When I was younger, I would go picking them with my dad in the fields near our house. When we got home, Dad washed them, peeled them, and boiled them in milk. Yes, milk! He used to make mushroom soup with milk, and this turned my stomach and I thought how can anyone eat this? I always thought it smelled like shit. I have seemed to have gained quite the audience in my crusade against mushrooms that I think now I have got to keep it going! I learned how to say 'no mushroom' in Chinese the moment I arrived there.

I can honestly say I have really come a long way, in terms of my eating habits and choices, since I have moved here. I am very much open to trying different dishes and cooking styles.

Coming from my previous job in retail, which had a very strict code of food hygiene and safety, I had a lot of food standards that I followed, i.e., people wearing gloves, hairnets, aprons and clean as you go systems. I have realised that food safety policies are not as strongly enforced in some countries as others. Sometimes you have to go with it here and just try things. Again, one thing to remember that if you are eating out and are going to try something different, bring a packet of tissues or a roll of toilet paper! Or if you go to Sri Lanka, bring your own knife and fork otherwise you should follow the locals and eat with your hands!

Customs

Again, as previously mentioned in an earlier chapter, the first time I truly experienced culture shock in terms of customs, was when I was in Jordan back in 2015. I landed a day earlier, which meant I had a day to explore the city of Amman before my group tour commenced. After negotiating a day out sightseeing with my taxi driver. I began the day's tour with a walking tour of the markets and bazaars.

As I walked down the narrow and ancient streets, my view was drawn to the hundreds of people wearing the traditional modest clothing that Muslims wear, with the men wearing the keffiyeh and ghutra and the women wearing the hijab, burka and the niqab. I was never exposed to this before in my life and at first it felt a little intimidating as I was the only foreigner in sight. I was not able to see another one in the area where I was and, given that there had been a terrorist incident in the previous months, and with neighbouring Arab countries battling terrorist groups, I felt slightly uncomfortable.

Then I realised that these attacks could happen anywhere in the world and that I did not have to be here to experience this. I could easily be walking down the street in Kilkenny and get knifed. So, to be honest I did not really feel too scared or at risk here. These countries do not really see tourists as threats but maybe some would to get more attention and to get what they want.

After buying some fruit in the market we were walking, when suddenly on speakers came the call to prayer. Arabic prayers came over large speaks throughout the city as some Muslims went down to their knees and prayed on the streets and some went to the local mosque which was across from where I was standing. Never in my life did I experience something like this, it was beautiful to witness.

Life stopped dead in its tracks, the traders and merchants stopped and dropped to their knees to pray. There was a point where I was standing in the middle of them all. The only one along with my guide and he explained so much to me about his religion, I was so intrigued by it. He brought me to the entrance of the mosque to observe the people in prayer but told me that non-Muslims were not permitted to enter the mosque. Just to stand there, watching and listening to the people pray was one of the highlights of that trip and a moment that I still think about.

Another experience of religious culture shock happened to me when I visited Jerusalem in 2016. This was by far the greatest city I ever visited and an experience that I will never forget. Visiting Israel was a dream of mine that I had for years, I was so fortunate and grateful to see that dream come true. I dreamed of visiting the Dome of the Rock, the Western Wall, the Church of the Holy Sepulchre, the Garden of Gethsemane and many others. Visiting this magnificent and ancient country meant that I would without a doubt be exposed to the religious conflict and feud that has divided this city for centuries. This trip brought me to so many amazing places in Israel which included, Tel Aviv, Nazareth, Galilee, Haifa, the Golan Heights (which borders Syria) and Jerusalem, and to Palestine, in which I visited Jericho and Bethlehem.

Experiencing these magnificent and ancient places and the conflict first-hand drew me to two conclusions. Firstly, there will never be peace in this land between the three major religions, it will always be like this and that is extremely sad. Secondly, there are no innocent sides to this conflict, everyone is to blame for what has happened, the Jews, the

Muslims and the Christians are all responsible for what has happened in this land. Now that might be insensitive of me to state, but this is a fact from someone who has studied this country and visited first-hand, as I mentioned in an earlier chapter. If you want to find out the truth about somewhere, go visit for yourself, it's the only way to do it. Do not listen to others, especially those who have never left their hometown. They cannot understand or learn from first-hand experience what life is like here because all they know is provided by the news reports and papers.

While we were travelling to Palestine, we passed numerous checkpoints which were armed with soldiers who has RPG launchers and machine guns and at the sides of the roads I saw tanks.

Now, what did I experience here that shocked me? I was on a group tour here also, and on the last day of the trip, we made it to Jerusalem where we would explore the city and see the most important religious monuments to three major religions of the world. The first one we visited was the Dome of the Rock (Temple Mount, Al-Aqsa Mosque, the site where Mohammed ascended to heaven, this is the Islamic connection. In addition, it is the site where the Temple of Solomon was situated, where Abraham was going to sacrifice his son Isaac, this is the Jewish connection, and the temple where Jesus Christ spread the word of God, Christian connection). We had to visit the site extremely early in the morning because it was Friday, the Islamic holy day. First, we passed the Western Wall on our way to the Dome of the Rock, that was an incredible site to see.

After numerous security checks we walked upon the plateau, observing everything in sight. We were informed that there were always two hundred soldiers and police patrolling the area at any one time and that Jewish people were forbidden to step on this area. Our guide did inform us that from time to time, rabbis and other Jewish leaders had to meet with Islamic imams and leaders. For this to happen, Jews must be escorted by some police officers and soldiers across the dome plateau to the meeting point to guarantee their safety and security. We were told that if they entered this sanctuary alone, it would have deathly consequences for them. The morning that we were there on the plateau, there was a Jewish man meeting with the Muslims.

We were observing the dome and appreciating the grand architecture of it, then we heard chanting and shouting. I turned my head slowly to see what the commotion was, and there were Muslim men standing up and running over to an area, where a man was being escorted through. It was the Jew walking across for his meeting, and the Muslim men were roaring and shouting at him "Allah Akbar, Allah Akbar, Allah Akbar." I felt a cold shiver down my spine and goose bumps on my skin, as this was perhaps the first-hand experience I had of the conflict in action, because I saw evidence of it in other places, but here it was in motion, happening before my very eyes. I can still hear those words being chanted patriotically and proudly. They were embedded in my mind, and I will never forget the sound of them.

It was this moment that made me come to the controversial conclusion that there will never be peace here and that everyone is responsible for the fine mess that has transpired here. "Allah Akbar" means "God is great" in Arabic, but when it comes down to it, these three religions all worship and believe in the same God.

Mannerisms

Different countries, different people all have different ways of life and mannerisms. What some people regard as normal in one country is frowned or looked down upon in another. The world is a great and amazing place. I have seen so many things, cultural norms and practices that have both amazed me and shocked me.

By far the biggest difference for me is the squat toilet! In Ireland and all western countries, we have the normal sit-down toilet, there are no squat toilets. The closest thing I have ever seen was the potty which little kids are trained to go to the toilet on. I saw a video on Facebook before I moved to China which showed the health benefits of using a squat toilet and I did think to myself, "Oh, people in the Middle East shit with great health benefits." Now, I only ever imagined that these toilets existed in middle eastern countries as I saw them in Jordan, but never expected them in China!

When I think of it, back in 2011 when I was flying to Sydney, I had a layover in Kuala Lumpur airport for a few hours, and I needed to use the toilet. While in there I saw a sign, which showed people how to use the toilet correctly! There were two people, one was sitting correctly on the toilet and the other one was a picture of a man squatting on the toilet, sure I was laughing at it and even took photos of it. I had never seen anything like that before in my life!

Then I landed in China. What an eye opener that was (and still is). So, after my first day in China, extremely jetlagged, we went to a rooftop bar for a few drinks, and this was a chance for me to meet some of the other teachers there. I had to get up and go to the bathroom, as I stood in a cubicle, I looked down on the ground and saw this opening. I just stared at it for about two minutes thinking to myself, *'how the fuck am I supposed to use this?'* When I got back to the table with the other teachers, I asked them the question, how am I supposed to use this thing? They laughed and said, *"Yeah, you are going to have fun trying to figure out that one,"* like I can understand what to do when you need to piss, but the other thing would take practice to master! There was some relief for me, as I was told that Starbucks in Guiyang had western toilets and that I could just go use them there. But I was not able to run around looking for a Starbucks every time I needed to use a toilet.

So, after a period avoiding using them, I decided that it was time to grab the bull by the horns and do it.

I will not go into details, but my attempts included me taking my shoes, pants and boxer shorts off and standing on my shoes, while holding the handle of the door as I tried to squat down! I did this because I was worried that I would fall into the hole! I shit you not! When I was walking out of the cubicle with my shoes under my arm, my colleague's wife was staring confusingly at me and asked what I was doing. I told her I was trying to learn how to use a squat toilet, she just looked and laughed at me. Six years later I still have not mastered how to use the squat toilet!

It is a common practice that grandchildren are reared with their grandparents for a good part of their junior years. In fact, many of the children whom I taught at my old school were brought to the school by their grandparents, not their parents. I guessed that this was because the parents are working for most of the day and trying to provide a good life for their child, so therefore they must sacrifice some time with them and let their parents raise their children.

I believe that it can be an extremely hard thing to do, but it is great to see the grandparents play such a huge and important role in the lives of their grandchildren. I know in Ireland that grandchildren visit their grandparents, and they spend time with them, but not to this extent.

My grandmother looked after me when I was born because my parents were working, and my brother was being taken care of by a childminder. My parents could only afford one minder at the time. So, I was given to my nan, but I was collected every day and brought home to my parents. I never once stayed over in my grandmother's house.

These are just some cultural differences, and we must accept them. Even though we might not agree with them, we must learn to go with them and respect them. As I mentioned in an earlier chapter, we got to shut up and get on with it. This is not our country, and we must respect it. At the end of the day, we are guests and workers in another country. If you do not like it here, a taxi can take you to the airport, it is about fifteen minutes from the city centre.

Another thing that I noticed here is that birthdays are much different here in China.

It is the cultural norm here that if you are having a birthday party or dinner, that the person whose birthday it is, pays for the meal or the party. Now in other countries, it is usually the person who is celebrating their birthday, gets treated to everything, and they do not pay for it, but here in China it is the opposite practice.

I was unaware of this until my own birthday on which I invited around twenty people for food to a restaurant. My friend asked me was I aware of the Chinese tradition that the birthday celebrant pays for their guests when they have a birthday dinner or event. I was totally unaware of this and thought it was great. I was more than happy to pay for everyone, because I would not want to be impolite or appear to be mean or anything, because I have so many good Chinese friends, I did not think twice about paying for them.

Because I know when it is their turn, they do the same, but I wish other foreigners would do that too. Chinese people are extremely welcoming and hospitable in which they invite you to eat dinner or to drink with them and if this is your first-time meeting with them, they may pay for it, and the next time, you pay for it. This is a good arrangement instead of going Dutch like most foreigners do, it saves disputes over splitting bills. But some foreigners I have seen take advantage of this in an unbelievably bad way. I have seen foreigners drink and eat with Chinese people, and when it comes to their turn, they let them pay for it again and again. In Chinese culture, this is seen as being 'mean' and word quickly spreads around, so the invite is not handed out the next time. I do not like people who take advantage of other's kindness.

At the Dome of the Rock (Temple Mount), Jerusalem

The Western Wall, Jerusalem

At the Stone of Unction
(The location where the body of Jesus was anointed and prepared for
burial), The Church of the Holy Sepulchre, Jerusalem

Walking at the West Bank Barrier, Bethlehem, Palestine

Looking over Jerusalem from the Mount of Olives

My first photograph in Guiyang

Chapter Thirteen

Through My Eyes

One thing I know for certain, if you do not try something, you are certain to fail. Now failure is a word that I have learned to hate. I refer to it as the *'f word'*. I tell my students not to use it, because it brands them and brings bad feeling to them. Before I get into this in detail, I will share with you briefly my experience of this.

As mentioned before, I went to Australia a few years ago in terms of work and a new life. I spent three months there in total and it was then I realised that it was not for me. After various job refusals and unanswered emails, I got very unmotivated about it and my life. I watched my hopes and dreams disintegrate. After some deep soul searching and decision making, I decided that Australia was not for me and that I wanted to return home.

As mentioned also in an earlier chapter that I came home because of my grand-aunt's health condition, but also realising that Australia was not working for me, also was a key factor in making that decision. As I flew back on the plane, I began to think, *'Am I a failure?'* and *'I could have done better'*.

Then I realised, quite simply, I had tried and after experiencing that, it was not for me. I did not feel like a failure, nor did I feel I could have done better in my circumstances, but I did it. I discovered some things are not for some people. I pondered the next day or two what would happen, would I find a job? Would I have to sign on for social benefits? What would I do?

To answer that question, within five days of returning home I secured a job in a supermarket. I picked myself up and I continued from where I left off. But I continued with life knowing that I had tried something new and even though it did not go as I had planned; I was extremely grateful for the experience and the opportunity.

Try. I tried and I knew that I did not fail.

Crusaders

Now, just where does all this fit in with China? One of the many things I have learned and experienced here is that quite simply it is not for everyone. People come here for different reasons such as work, travel, and the lifestyle. Now, these are great reasons why they would come here. Some have come here and settled into life in China. It is not an easy transition to do but they do it. Some try and settle into it but for some people they cannot complete the transition and therefore, they falsely do it, and there are also those who refuse to adapt to the surroundings and culture.

For me, it took time to adjust to the culture and to the ways of life living in China. I worked at it, and I reaped the benefits. I adjusted to the food, the chopsticks but not the squat toilets! You must realise this is not a western country and you cannot expect that everything will work as you would want it to. In addition to this, you cannot go around trying to teach local people on what is right and what is wrong. This is something I see so often. Some westerners come here with principles and ideas that they can educate people here and 'do good' by telling them to stop doing certain things. It cannot work. It is a waste of time. End of story!

In China, you must learn to go with it and sometimes you are going to see things that will make your skin crawl or even upset you. I was told when I arrived here back in 2016, that you must hit the 'reset' button inside your brain. Remember this, you are in a different country, a different culture and experiencing a different way of life. Political and social correctness is not accepted in many places.

Some people might feel they are contributing and educating people how to live, what is right and what is wrong, they embark on a crusade. But guess what? It is not going to work. Who are we to go to another country and try to make people change? It is very patronising and insulting to people of different cultures to have some foreigners come and try to change things. Again, it is not going to work, we are guests in their countries, ambassadors of our own lands. Therefore, whether we like it or not, we must respect and go with it otherwise you are not in the right place. Simple as that.

The Wrong Reasons

Some people come here for the wrong reasons. I will explain this by going back to when I was in university. Back in 2008, when I started studying, my mother always told me that there were two types of people. The first type were there to get an education, to work and get a degree, this was their focus. The second type of people were there to enjoy the lifestyle and the freedom. She would also speak of the latter boasting, "I'm in college and I am enjoying the life," but the purpose, the 'why', was wrong. These people would fail their exams therefore wasting years of effort (mostly minimum), money and even some of them would drop out before they completed their time.

More than likely, these people were at university to escape home. Their first experience of life away from their parents and their surroundings so even though they are in university, they are there for the wrong purpose. Drinking and partying every night, I have seen students get into fights, trash university and private property, and inevitably, drop out, because ultimately, they went there for the wrong reasons.

You must also consider the extreme costs of going to university, the financial strain it puts on families, the sacrifices parents make so their child can have the opportunity that perhaps they never got. What do they do with this opportunity?

They piss it down the toilet. So many people want and need this opportunity, but they cannot afford it and it's a shame. So many people are refused financial support and grants, who genuinely need them, and then you have the people who squander it.

While all the time, it could have been used to serve people whose 'why' was greater and more justifiable.

I have seen people come here to teach, but their motivation and reason is not to teach. Rather, use it to learn Chinese and use the experience to focus on the culture, not the work. Now, it is perfectly fine to do both, learn the language and the culture, but it is not fine if you do not make the effort in your position as a foreign teacher.

They maybe be amazing at the language and have been to all these beautiful places, but that is not what they were hired to do. Some people have got lazy and failed to fulfil their jobs as teachers and therefore have become alcoholics, which I have seen. Refusing to change their ways and teaching styles. Others doing half their jobs, coming in hungover and putting minimal effort into their jobs. I have met my fair share of people like this here over the years.

I heard someone say that 'paper is patient'. Everything looks good on paper but really, 'the proof of the pudding is in the eating of it'. I use these metaphors to describe teachers, to really see what they are like and if they are suited. It's not through amazing interviews or through beautiful and artistic resumés, but rather by putting them in the classroom and seeing if they can cope with it. I say cope because it is a little tough. I stated earlier in a previous chapter you must be prepared to change your ideas and methods quickly if what you are doing in the classroom does not work.

Money is not everything

Hollywood Actress Bette Davis said:

"To fulfil a dream, to be allowed sweat over lonely labour, is the meat and potatoes of life, the money is just the gravy. As everyone else, I love to dunk my crust in it, but alone, it is not a diet designed to keep body and soul together."

I was about twenty when I first heard this quote. Davis summed it up, that it is not just about the money, even though it is great, it should not be the reason why you do something. For me, doing something that you love should be the biggest reward, the best feeling, not the money.

Something else that I am noticing here from my time living and working in China, is that you will meet people who will feel they are being 'hard-done-by'. They feel that they are working too much, and they feel they should not and that it is not fair. Sometimes people's motivations are different, they favour the lifestyle more than the reason why they were employed. They demand a certain amount of pay and attempt to dictate their own working times. People have such an attitude of self-entitlement. It pisses me off.

They were not employed to work with little or no effort. They were hired as teachers who must give 100% to their students and to the school,

Now, sometimes it is difficult to do the latter because of policies but regardless, it is the school, the manager or the owner who pays your salary, and it is a decent salary and it also has many other benefits, housing allowance or house provided, free meals, utility bills, medical insurance and flight ticket reimbursements are also provided, not to mention a contract renewal bonus. For the hours some work, some would argue they are underpaid, and some would argue that they are overpaid. Sounds crazy but it is true, who would ever think they are being overpaid? I think the type of people who feel they are overpaid are those who do not do this job for the money, but rather feel they are being given the opportunity to grow, learn and to pass something on.

On the other hand, there are many who feel they do not get enough for what they do. Different schools will dictate different expectations of teachers, depending on their qualifications, and therefore teachers will get paid different salaries.

What we do, what ESL teachers do is not strenuous work, it can get tiring, but it is not stressful, nor is it a tough job, well for me anyway. Working eighty plus hours a week in retail management is tough, that was my life. When I heard stories of my dad, my uncle and even my grandfather going out into the fields tinning beet at the age of twelve or fourteen for eight, maybe ten hours for a few schillings (which was not much at that time) for the day's work. Now that is what you call labour, tiring and hard work. My mother's parents in London, working their asses off for a better life. I am eternally grateful that the values of hard work were instilled in me, and it has made me appreciate everything I have. Not given, but earned, this is what my family taught me.

In my opinion, teaching is not hard work. Now, there are aspects of it which can be difficult such as dealing with difficult kids, trying to manage a high energy level class, and dealing with a difficult parent. Now, that can be hard work but all in all, it's not that bad. It's only hard work if you make it hard work.

There are so many people who wish they could do what foreign teachers do. People who have dreams and sadly never got the chance to pursue them due to events and reasons that are out of their control. Then there are people on the other hand, who do not 'fit' in the role of an ESL teacher.

Now if you, the reader, feels in some way inspired to pursue a career in ESL teaching abroad in China or anywhere else, please consider the following few points that I am about to share with you.

Firstly, do not do it simply for the money. This never works. If money is your goal, go work in the mines in Western Australia where perhaps your only responsibility is to operate a machine for hours, days and weeks on end and watch the piggy bank expand. Do not do ESL solely for financial gain. I will admit that it is a great way to save money, but soon enough, in the classroom you will get tired of it, and you will begin to question yourself why you are doing it. It is not all about the money.

You will find it exceedingly difficult to embrace the culture and make friends here if it is all about the money. I see this too much. So, do not waste your time coming here to strike it rich.

Secondly, do not do it to 'find yourself' or simply to 'learn the language' because you will find that you will be required and expected to do the opposite. If you want to find yourself, go to a monastery in Thailand or help the poor in Africa. Sometimes, people who just want to learn the language, forget what they come here for and instead of teaching English.

Do it because you want to experience life outside your comfort zone, do it because you want to walk that path that very few have the guts to do. Do it for the chance to experience different people and different cultures. Yes, it is difficult when you cannot communicate with people when you first arrive, and when you want to eat chicken in a restaurant and you cannot say it, and therefore you must use charades and body language to get what you are looking for. Be ignorant to everything new that is around you, treat your mind like a new slate, untouched and ignorant, taking it all in, embracing it and learning it. This is what I did.

Do it because you feel you can contribute, inspire, and help others in their lives. People always remember that one teacher or that one person who helped them learn. You can be that person. From my experience, Chinese students appreciate having a foreign teacher and they can help the students do better. They are motivated and inspired by them to get through their studies. I have had the honour of teaching so many teenagers at my current school and they have also inspired me to become a better teacher.

I got home extremely late one night during Christmas 2017, and I was going to bed about 11.30 pm and I noticed a message on my phone from who I thought was one of my students, but in fact it was his mother asking me to help her son. She felt that his focus was slipping and that she was scared his grades would fall and that he would not be able to get into a good high school. She turned to me for help. She told me that I am an inspiration to him, a role model and that I am his favourite teacher.

I was really taken back at this, because I did not see the positive impact I had, that spending two hours a week with kids can have. I felt so honoured and so proud that I was asked to help her son. A few months prior to this, I had read a book called *The One Thing* by Gary Kellar and Jay Papasan. It is my favourite book. I have bought several copies of it for friends both in English and in Chinese. I decided that I would focus my lesson on the principles of *The One Thing*. My objective for this class was to motivate the students to focus more on their studies and life ambitions. The following day, I received another message from the mother expressing her sincere thanks for inspiring her son and that he was more focused and determined. Over the next few weeks, he rang me a few times to get some advice on his exams and his approach to his study. I reminded him of his goal, to get into the No. 1 High School in the province. I told him to bear in mind the goal and to keep his focus on achieving this. A few weeks later, I got a message from him telling me that he got into the best high school in the province of Guizhou. This is also the school in which I am currently teaching, and I was so lucky to have two years with him in this school, where we would regularly meet and chat. He has since completed school and is currently a first-year university student. I am so proud of him.

It was from answering this mother's call for help and helping her son get into the school. I realised my worth and that I am doing what I was meant to do in life. Help others. A close friend here told me, "Helping others is helping yourself".

I found my calling.

Chapter Fourteen

Take the Good with the Bad

Why do we travel? I guess that is an easy question to answer. We can say that we travel to different places to experience different cultures, see new things, meet new people and try new food. I could talk for hours answering this question. I have narrowed it down to just one word. Experience. That is my main objective of travelling. Everything we do in life is an experience. It cannot be taken away from us no matter what happens.

Now generally, we travel to have a good experience, but my own personal experience is that not every experience can be positive or good. There must be a mixture of happy and sad experiences. If we were to have a happy experience every time we travel somewhere, to be honest it would make it quite boring and repetitive. Perhaps, one would even get tired of it quite quickly. As you will read in this chapter about my experience in Cambodia, you will learn that we need to have both good and bad experiences. This chapter will focus on some eye-opening experiences which helped me appreciate all the good ones I have had in the past. It made me have a different perspective on travel.

One evening in the Summer of 2001, my mom took me to the cinema to watch *Lara Croft: Tomb Raider*. I had been (and still am) a noticeably big fan of the video games for over twenty years, and now they had brought the videogame heroine to the big screen. I was always intrigued by these video games; each game took you to a different world and a different culture. For me it fed my curiosity of ancient history and antiquity. Little did I think playing video games would inspire me so much.

The plot of the movie was Lara attempting to stop the Illuminati from gaining control over time by using an ancient artefact which gave the possessor the ability to control time. The first stop on her mission to stop them was to the ancient temple complex of Angkor Wat in Cambodia. I was mesmerised by such a magnificent site. Sixteen years later, I would visit the temple complex.

In the Spring of 2017, I had a three-week vacation for Chinese New Year. I had just spent ten days travelling around China with my friend Stephen, who came to visit me for two weeks from Australia. After visiting so many of the biggest monuments and attractions in cities such as Chengdu, Beijing, and Xi'an, I was about to jet off to Cambodia for a week's adventure in the sun.

I was excited to go to Cambodia as I was going to visit the famous temple complex of Angkor Wat and I was going to enjoy some good weather after bracing minus freezing temperatures in China. Upon arriving in Siem Reap, I met Raquel, one of my friends from Guiyang. She had flown in from Vietnam for two days to explore Angkor Wat with me. We ate dinner and planned to explore the temples at sunrise, as it is 'one of the must-do-things' in Cambodia. So, we spent most of the following day exploring the temple ruins. We had our 'Tomb Raider' on, as we posed for so many photos and explored like Indiana Jones or Lara Croft would have. As we were walking through the Angkor jungle,

Raquel noticed some kids sitting and playing on a wall. She saw this as the perfect photo opportunity to capture. As she took out her camera, the kids stood up on the wall and said, "Lady, you take photo, you give us five dollars." We just looked in surprise, first at the kids and then at each other. We could not believe that the kids who must be as young as five or six would not just say this but also know the meaning of this. I guess with the number of tourists who pass through the temples and jungle, the temptation is there for everyone to try extorting what they could from foreign tourists.

As we walked past, we started discussing this in depth a bit more. We were shocked, but as we walked on, we met a small boy carrying a huge basket filled with magnets and souvenirs. The poor child's arms were heavy and burdened by this basket. He limped past us as he asked us to buy from him. Something that I learned later about this, whether it is true or not, but Angkor Wat has tourist police who monitor the ruins to ensure no crime is committed and that these children work for the police officers to sell souvenirs and trinkets for them. Now, this in my opinion is child slavery. But in the eyes of a Cambodian, it might be a child trying to work to support a family that has little or no income.

Little did I think that this story and experience was just the beginning of several horror stories I was about to learn about Cambodia. This experience is not designed to put you off going to Cambodia, but it reflects someone who went into the country completely blindfolded and ignorant to the reality of the country. After my friend Raquel returned to Vietnam, I was enjoying my dinner in the famous Pub Street area of Siem Reap. It was the place where I tasted tarantulas, locusts, snakes, and scorpions. Talk about eating crazy things, I would eat things like this, but

I never (still never) have eaten a mushroom! While eating my dinner, I was chatting with some other tourists, from Germany and Singapore, who had been enjoying their time in Cambodia. We started to chat about the darker side of Cambodia. I told them about what I had experienced at the temples, this was nothing compared to what I was told by them.

They told me that they heard stories of families who are so much in debt and are so poor that the only way to repay and survive is to send their children out to 'work'. Work meant selling their bodies to pay for food. One fellow traveller told me that he had heard a story where one family was hounded by some loan sharks because the family was unable to repay their debts, so they took the daughter of this family, just fourteen years old and gang raped her. Shocked was not the word to describe my feelings about this. I guess that when you are at home in the comfort of your own home and surroundings, you forget that some people are not as fortunate as us.

We chatted late into the evening, and I told them that I was planning to visit Phnom Penh the next day. I had already booked my bus ticket with the hotel. They asked me what the bus company was named, and they told me that I picked the worst company. I asked them why and I was informed that they have a habit of 'losing' passenger's personal belongings and robbing people on the bus. I was very worried, and I could feel my anxiety starting to flare up. On top of this, I was also informed about just how unsafe and dangerous Phnom Penh was. People get robbed from the tuk-tuk vehicles by passers-by who just grab handbags and backpacks and there is nothing that could be done about it.

After I returned to my hotel, I spoke to the concierge about my concerns. She had been so nice and helpful to me, I decided to get her opinion on what I had just learned. She told me to exercise caution especially in Phnom Penh, not to carry much cash on me, to leave all valuable items at the hotel and not to take your camera or phone out in public. As for the bus company I had booked with, she refused to offer me any advice on it.

The next morning, at 8 am, I embarked on the eight-hour bus ride to Phnom Penh. What I experienced in Siem Reap proved to be relatively mild to what I was about to experience in Phnom Penh.

Before I get into talking about my experience in Phnom Penh, I want to share with you, the reader, about my ignorance on what I was going to learn about here.

To cut a long story short, I had never learned or read about the terrible history of Cambodia and the Khmer Rouge. I never realised just how many people died there and how cruel the Khmer Rouge leader, Pol Pot was. I really did not do any more research on this country apart from the temples at Angkor Wat.

I arrived in Phnom Penh late afternoon, and I had spent some of the journey talking to a New Zealander who was on holiday there. I almost mistook him for American actor James Brolin, married to Barbra Streisand, but the New Zealand accent and his 25-year-old girlfriend proved me wrong! He was an English teacher in Brunei, so it was interesting to learn about this country. Brunei had always been a country that I wanted to visit, as it always seemed mysterious to me. I had the chance to visit this country in 2018.

Up to this point, I never met anyone who had visited Brunei before. I asked him about what I should visit and do in Phnom Penh. He recommended the Killing Fields, museums, the royal palace, and temples. He told me that the genocide museum and the Killing Fields were not for the faint hearted. He said no more. I think he realised that I did not know anything about them and was not about to spoil the experience.

Now, when I initially heard Killing Fields, I immediately thought they were like the battle fields of Normandy, where battles took place. It was not until I searched on the internet about them that I began to realise the extent of suffering and turmoil Cambodia went through. I was about to experience something that would have an everlasting impact on me.

The following morning, after eating breakfast I headed off in a tuk-tuk to Choeung Ek, the best-known Killing Field. Upon arrival to this place, I was given an audio headset at the admission office and a map. As I entered the gates, I saw a tall tower, a white steeple in the centre of the area. As I followed the map to the first point of interest, I put on my headphones and turned my audio guide on. The horrific details began filling my ears and mind with the ruthless re-telling of the barbaric torture and killing of over 1.7 million people between the years of 1974–1979. Whatever joy and happy experience I had of Cambodia up until this point was swiftly leaving me. I stood in shock among countless other tourists who were also learning the truth.

First, I was introduced to Pol Pot and the Khmer Rouge and what they stood for. Passing down the stone grassed paths, I noticed huts with straw and wooden roofs covering areas of land. I was not sure what these areas were, but then I learned that they were mass gravesites that were the resting place of numerous victims. There was an information plaque which read: *'Mass grave of more than 100 victims, children and women whose majority were naked'*. This sent shivers down my spine as I read it and I looked at the earth. I could see bones sticking out of the clay, this was due to heavy rain, the bones would wash up from the ground. It goes to show just how shallow the graves were.

I couldn't help but think of the mass graves at home in Ireland from the time of the Potato Famine (1845–1849) and how when I visited the famine graveyard outside my hometown, I felt sad as I thought of the number of victims who were buried there in unmarked graves. I had that feeling as I walked through the Killing Fields.

As I walked on down another path, which was shaded by trees which I thought were banana trees, the leaves and bark had razor sharp edges like teeth. I was informed that the prisoners were shoved against them, and the impact would slit their throats without a cry or a shout. I touched one of the razor-sharp leaves and I could feel the jagged edge and could imagine what those people went through. One could say that everywhere and everything you looked at here was both an area and instrument of torture, pain, and suffering.

Perhaps the most vivid image that still haunts me to this day was a tree. I looked at the information sign beside it which read: *'Killing tree against which executioners beat children'*. When I think of all the different ways that we can use trees, never in my life could I imagine that they would be used as such a barbaric torturing device. I didn't need the assistance of the audio-guide to tell me how they killed small children and babies, but to imagine that the executioners would grab children by their legs and bash them against the tree and then throw them into a mass grave, is pure evil. Innocent and harmless children who have no part in this. Then, their poor mothers were raped and then met a similar faith.

The second last part of my visit to the Killing Fields (brought me to another tree in which during the time held a radio and a megaphone. Though this radio and megaphone uprising and rebellious speeches and songs were broadcasted. This gave the impression to local events that rallies were being held here, but in fact, they were used to drown out the slaughter and the cries of the victims of the Khmer Rouge. A loud generator also accompanied the music which gave power to this hell. As the guide finished talking, I could hear the songs and the generator playing on my headset. I will never forget the music, nor the sound of the generator, they have left everlasting impressions etched on my mind. They will never leave me. I can still hear that music as I write this.

The final part of this horrific experience was visiting that tower that I mentioned upon my arrival to the Killing Fields. As I approached it, it got clearer to me what this tower was used for. This was a tower which had eighteen shelves in it, and each shelf contained human skulls. Each skull on each shelf had a different colour dot on it which had me thinking about what it meant. But it soon became clear from reading a sign that these dots indicated how that person was killed. Whether it meant being hacked with a knife or a long thin blade drilled into the head, sadly it allowed me to visualise the barbaric torture and suffering.

Placed upon the shelves alongside the skulls were the instruments and tools of torture which were used to carry out the executions. As I walked away from this horrific shrine, I purchased some flowers, laid them by the tower and bowed. I had two reasons for doing this. The first was for sympathy for the victims of the Killing Fields, and second for my own ignorance.

As I mentioned previously, I did not know anything of the history of Cambodia. I wasn't aware of what happened here but by God was I slapped in the face when I began to learn about these places. Outside the tower, located on the wall was a sign which read: '*Would you kindly show your respect to many million people who were killed under the genocidal Pol Pot Regime*'.

As I left the Killing Fields, I walked out and did not turn back for one last look. This is something that I always do whenever I visit a site or a landmark, one final look to etch it on my mind. There was no need to do that here, as what I felt, and saw would have an everlasting mark on me.

While I was on my journey back, I cast my mind back to when I was in university and when I was studying a module called tourism studies. While studying this, I remembered I completed a project on 'dark tourism' or 'thana-tourism'. My research showed the feelings and reasons why people would visit places like the Killing Fields or Auschwitz Concentration Camp. Now after eight years of doing that research, I finally could see those feelings and reasons.

As humans, we tend to get thrills from all kinds of experiences, and there are some who get their thrill from visiting such places as mentioned above. This was something that I could see at the Killing Fields. But I then saw the other experience. I saw local Cambodians walking around this place with their head down standing and reflecting at the mass graves. I could think that perhaps they had loved ones who perished here at the hands of the Khmer Rouge and that their bodies were thrown into the mass graves, unnamed and unmarked. I could only imagine the people visiting the steeple with the skulls walking past thinking 'is this my father, my mother, my son, or my daughter?' These are the questions that so many Cambodians will forever ask and will never be answered. Perhaps that skull staring back is that person's family member. We will never know.

The next stop on my tour that day was to Tuol Sleng Genocide Museum, or better known as Street 113. As if I was not satisfied with my genocidal experiences, I was about to go through it all again. Again, I was guided through the three-storey buildings and courtyard by my audio tour guide. These buildings consisted of torture and interrogation rooms, while some still had tools and instruments of torture inside. As I walked around the rooms, the walls and floors were covered in stains. On one floor there was a message that read, *'We didn't love each other, but when Angkor asked me to marry, I pretended to follow the order, so I could survive the next moment.'*

It seemed that people would do anything to survive here, whether it be for an hour or a day. I also imagined that the people had to carry out acts against their will just to see the light of day for just one more moment. I continued and went into another building which was inside a room. The walls were covered with photographs of Pol Pot and the brothers of the Khmer Rouge. Then inside another room, there were thousands of photographs of the people who were brought here.

From young to old, their mugshot images with their facial expressions posing the question, "What did we do to deserve this?" It shocked me further to learn that people who wore glasses were condemned because they were seen as being educated.

As I walked around the room and passed the photographs, I observed other visitors looking and studying the photographs, perhaps asking, "Where are you now?", "What happened to you?" I looked at the tears in their eyes which made me wonder that perhaps they were family members of those who were taken. I spent some time roaming around the torture rooms there, and after a short while I realised that I had experienced enough for one day.

The next day, I took a ferry to an island which is renowned for making silk. I thought that it would be interesting to learn more about this trade, as living in China, I learned about the silk route, but given Guiyang's location I am far away from where it begins. I spent a couple of hours travelling to different parts of the island learning how they make silk and even got the chance to take part in the weaving process. I then purchased two handmade silk scarves, one for my mother and one for my grandmother. I picked up a third one accidentally for my grand-aunt Marie, but then I realised that she was no longer with us. It had become a habit of mine for so long, to always buy her some souvenirs, even after losing her four years prior, some habits are hard to break.

After I returned to my hotel, I changed my clothes and went out to eat dinner. I went to a Middle Eastern restaurant where I had the chance to enjoy falafel and hummus. Middle eastern food is one of my favourite cuisines, and you would not believe how much I enjoyed this, as the last time I enjoyed it was when I was travelling around Israel in 2016.

While eating I engaged in people watching, an activity that I enjoy doing with my uncle Sean whenever I visit him in England. We would walk down the street, buy newspapers and some pastries, and sit on a bench and watch the world go by and comment on what we saw. Now, people watching in England was totally different to people watching in Cambodia. You see some things that you wish you would never see. You see some extremely uncomfortable and disturbing things that bring a lot of the fears and problems in the world to life.

I noticed a young boy, who could not have been any older than five years of age pushing a little bicycle around the front of a small convenience store beside the restaurant where I was dining with no clothes on him, not even underwear. He was quite content pushing his bicycle around, probably one of the few possessions that he had. Then suddenly I saw a foreigner walk up to him and take out his camera and point it at him. Immediately, the boy turned away and covered himself to stop the man taking photos.

What happened next really shocked me. The man bought him an ice-cream and handed it to him along with a few dollar bills, and then to my horror, the little boy turned around and exposed himself to the man to let him take photographs. I sat in utter shock at what I just witnessed. The thoughts that filled my head were 'Was this man a journalist who was highlighting the impoverished and desperate ways of the Cambodian people and what they do to survive day by day?' or was he collecting images of young children for his own disgusting pleasure.

Regardless of the reason for this, it was a horrible thing to experience. For me, it just highlighted the incredible corruption and despair that had taken over this country and the desperate measures people and children are subjected to take to make a living. Even today, I think about the poor children and people who must subject themselves to this at this very moment to help their families survive, or perhaps they don't have families and they are left to look after themselves.

To summarise my experience in Cambodia, it provided me with so much knowledge about the culture both good and bad. I am profoundly grateful that Cambodia taught me that not every experience will be positive. I went to Cambodia ignorant and left it educated about many different things. To sum up this whole experience in a few words, 'experience is the best teacher'.

The Great Wall, Beijing, China

Admiring the temples, Angkor Wat

Paying my respects to the victims of the Cambodian genocide

Chapter Fifteen

Goin' Back

In August 2018, I sat at the Hamad International Airport in Doha, Qatar, waiting for my connecting flight to Dublin. One year since I was with my family and friends. One year since I hugged my parents, held my grandmother's hand, played with my dog, cut the grass, slept in my own bed, and walked the streets of the town and city that I knew so well.

I had some strange feelings about going back home after a year. I really wanted to go but at the same time I had a feeling that I wanted to go somewhere else, like Sri Lanka or Myanmar, but the fact that it was so long since I saw my parents and my grandmother, I wanted to spend time with them in Ireland and visit my aunts and uncles in London.

A funny thing happened at the airport. I went to a coffee shop to get a takeaway drink while waiting. I ordered my drink, but I ordered it in Chinese! The barista just looked at me and grinned. It took me a few seconds to comprehend what I had just done, and when I did, I started laughing. I told the barista I had to change my language when I go home to Ireland.

My mind was still in Guiyang. While I was waiting for my flight, I wanted to put some more songs on to my phone, so I had more to listen to while on the journey. A song that I had a strong urge to listen to was '*Going Back*' by Dusty Springfield. This song summed up my feeling that I was returning to the place that I knew so well in my life, but at the same time I had grown and that things would not be the same as they used to be when I was there. Nothing can stay the same. Where there is no change or struggle there can be no growth or even appreciation.

This part of the journey did not take too long because I was knocked out for most of the flight. I had taken a strong Valium tablet and it really kicked in. As we prepared to land in Dublin, I could see the green fields and the small villages and towns. These things had been missing from my life over the last year. As we taxied on the runway, I saw the name of Dublin Airport in Irish and I thought to myself, 'Here we go.' I was excited to see everyone but at the same time it felt so strange.

I met my parents at the arrivals area, and it did not take long to get back home to my house. On the way, I stopped into the nursing home where my grandmother has been living for the last two and a half years. This was the only thing I wanted to do on the first day back, see my grandmother, because this year started off badly for her as we thought we were going to lose her due to a very bad cold.

I had been travelling to Brunei, Thailand, The Philippines and Singapore for the Chinese New Year and I got the call which I was told to be prepared for the worst. I had a flight picked out to get me back home and insisted all arrangements be put on hold until I get back. Thankfully, it did not have to come to that as she had a full recovery and is doing quite well now. I was so happy to see her. It was a reunion I did not think I would have the chance to have again.

The next few days that followed felt so strange to me. While I was at home in my house with my parents, I felt fine, everything was okay. When I went to my city, I started to feel this weird sensation. While walking around Kilkenny, my heart began to beat fast, and I felt out of place here.

Some things that I knew from a few years ago had disappeared, shops, cafés, and some people. I walked around and felt that I did not belong here. I met up with some old friends and some old work colleagues from my days when I worked in retail. Even though it felt good to be with them, it felt very awkward. Awkward in the sense I didn't know what to talk about, I didn't want to be always talking about China. Even though I enjoyed telling people of my experiences, I also wanted to know about them and their lives, what had been going on and what they were doing.

I went home later that day and my parents asked me what the matter was because they found me a little strange. I told them about how I was feeling. But I told them that this is something that they know nothing about, because they never really experienced this feeling because my dad had always been in Ireland and my mom had moved to Ireland from London in 1979. So, it was so hard for them to relate to these feelings of being overwhelmed by everyone and everything. I told them that it felt strange being here, that I did not know what anyone was doing and what was going on in their lives, and that everywhere was different. I felt like an outsider.

The next few days were no different. I went out one Sunday night with two school friends, both of whom also lived abroad teaching but in South Korea. They were able to relate to me on how I was feeling and that it was normal. I felt a bit better that it was not just me. It was a great relief to know that others had felt this way too. They both told me it takes time to re-adjust to life at home, even though I had just been away for one year, but it would take some time.

I was with two old friends but still it felt uncomfortable as it was ridiculously hard for us to find a mutual topic of discussion because we all had different interests. Their interests were in Irish sport because the All-Ireland Championships were on in the coming weeks, and I never showed any interest in that. I think I only saw about four or five of the games on television.

The next day, I went on a drive with my parents and again I was telling them how I felt. My mom asked me did I regret coming home. To which I replied and told her that I did not regret it, but it felt strange because we all get so caught up in our lives it can be hard and difficult to reconnect with some people. Over time the connection can become weaker, my mom agreed with this feeling. Finally, I felt understood.

I spent some time reflecting on my feelings and my situation and I concluded what was wrong with me. I was suffering from reverse culture shock I was the 'minority in the majority'. Having been surrounded by Chinese people and being the only foreigner around, where I was the minority, now I am taken out of that comfort zone and thrown back to my old situation, where I was part of the majority. I felt everything was different, the places and the people. Even my own brother, it was quite hard to find a mutual level to communicate with him on.

During the second week of my trip, I went to London to see my aunts and my uncles. As mentioned earlier in the book, it is so important for me to see these people every year. I started my journey in Southend-On-Sea, where my uncle Sean lives. I spent most of the trip there and used it as a base to travel on to see my aunt and uncle in Bromley and my aunt and uncle in Windsor. I was delighted that my uncle Sean was able to relate to me about the disconnection feeling. For Sean, who is now in his eighties, has spent the best part of sixty-five years of his life in England, so he said that it is normal to feel disconnected and feel like a stranger.

I could only imagine that if he were to return to Callan, some sixty-five years after he left, how difficult it would be for him to readjust to everything. Not just living here, but also how difficult it would be for him to find anyone who is still from his youth. I think there may be one or two around still, but I could only imagine how difficult this would be for him.

The next day, I went to visit my dad's younger sister Gabrielle and her husband Roy. Like Sean, Gabrielle had left for England and even though she had not been over there like Sean, who left Ireland when he was quite young, she still had the same feelings and understanding as me. Again, I was so relieved that it was not just me who suffered from this. I was not alone.

I then continued to Windsor, to my other aunt, Molly, and her husband John, who had spent many years living in South Africa. Now, they had lived in England prior to emigrating to Africa in the 1960s. They too felt it was a change to their lives and their routines, and that it took them a while to re-adjust back to life in a country where they had been absent for most of their lives. Now, I felt I was following in the footsteps of my aunts and uncles, who had left home to pursue a new life. It is just that you can get so used to it, the way of life in another country and that it can be quite hard or impossible to change back to the old way of life. It was so hard to imagine that only after two years away from Ireland, this was my feeling.

On my first visit back home to Ireland in April 2017, when I returned for a week to surprise my dad for his birthday, I was in the kitchen with my parents and my brother. My brother was telling me that I would be home in August and that I can concentrate on getting a job and settle back into life at home, to which I informed him that this was my life now and that the possibility of me ever coming back to settle in Ireland were indeed very low. My brother looked at me, with a crushed look, he was shocked by what I had said. I said to him I need to make my own decisions based on what I want in life and not what others want.

Now, it might appear that I was a little too harsh, but it is the truth, and the way I feel. I may never come back, but these are the decisions that I will make in my life. These are the decisions that each one of us should have the courage to make by ourselves and not let others influence us and make the decisions for us. Too many people let this happen to them, and as a result they live with envy, jealousy, and regret.

So, going back to my trip home, after I returned home to Ireland, I went off for two days down to the southwest of Ireland to the town of Killarney. A town where I used to spend my childhood holidays in. I wanted to spend time with my parents and thought that this would be the perfect place to do so, as this is the place they love so much. I would encourage anyone reading this who lives abroad to visit home, and even those living at home to take your parents away for a few days, spend time with those that are close to you, if you are lucky to still have them because it is the precious time that you will never get back and it is time you reflect on while you fly back to say goodbye to them. So, forget boozing for a few days on your trip home with people who do not really care about you or what you are doing in your life and spend it with the people who care the most about you.

I had a few priorities for my trip home, first to spend time with Mam and Dad, see my grandmother, visit family in England and to go on a trip to southwest Ireland. The reason why I wanted to go on this trip in Ireland was because for the last two years, I felt quite disconnected from my Irish identity and my culture.

It was wonderful to soak up the Irish atmosphere in a town that I loved. The food, the accents, the scenery, and the music gave my soul and my mind the replenishing that I needed so much. I never realised just how much I was disconnected from Ireland and just how much connected I had become with China. I felt that I had adopted a new culture and a new way of life from living here, something I know many expatriates are not able to do here in China. As I walked down the streets of Killarney listening to Irish music, I heard one song called *Willie McBride.* I stopped outside the bar to listen to it, and it made me reflect on everything that I have done, and it made me think about living abroad. It made me feel so proud of my life and that this is what I was meant to do with it.

As the final days of my trip were approaching rapidly, I spent them at home with my parents and limited my trips out to meet people as I wanted to just switch off from everything, forget everything for a few days and have a few days when I didn't have to tell my story. It was during these days that my parents were finally able to understand my feelings and how overwhelmed I was. They started to understand the term reverse culture shock and how it had affected me.

I think this trip proved to them that I was perhaps gone. Gone, for the meaning of settled in another country and that I had made a new life for myself and made so many wonderful friends. They are Irish, Chinese, Canadian, French, American, Chilean, English, and Scottish. These are wonderful people and very dear friends, some of whom I consider now as family. To be honest, I consider some of them to be more like family than some of my own relatives.

I was quite happy to return to Guiyang as I had so much to look forward to when I got back, seeing my friends, my students, getting back to the gym, to future trips in Asia and the food!

When I returned, it seemed like I was never away from Guiyang. On past trips, I felt like I suffered from PHD (post-holiday depression) but not this time, I felt great. It was wonderful to be home in Ireland with my family and my friends, but it was also wonderful to have something great to look forward to getting back to.

Life is wonderful.

With Mam and Dad, Ireland, 2018

Celebrating my birthday with friends

Conclusion

My goal for this book was to take you on the journey that brought me to where I am now. I started off in life as a home bird, not aware of what lay ahead for me out there in the world to discover. On the journey as you read, I faced many problems and issues from the death of family members, health scares and anxiety. These experiences have made me stronger, even though still today I carry the scars. However, the issues have served as teachers to me. I showed you that losing a loved one does not mean it is the end, but it can be the beginning of something new.

I have faced a lot of things in life, but I have risen stronger from them. These experiences have shaped me into the person I am today.

There are some points that I hope you will reflect on from my stories. Firstly, travelling is not always about getting the best selfie or doing it for Instagram, but rather taking time to reflect and appreciate other cultures and the hard times that they went through. You must understand that you will not rave about how amazing everything is, because if we did that, we would all get bored of travelling so in my opinion there must be a balance of both good and bad experiences. You must visit places like the Killing Fields in Cambodia or visit Auschwitz and understand what happened there.

Secondly, when you travel you must travel with an open mind. It is essential to have this when you go and experience a different culture, people of different beliefs, customs, orientations, and ideas. Yes, you will have culture shock and some things will remain puzzling to you, but you must go with it. You may cringe at some of the things that you see but you just must roll with it. I have seen so many things over the years that I wish I never saw, but then again, these practices and ways of life belong to another culture. You do not have to understand them, but you must respect them.

Thirdly, travel will teach you so much about yourself. You push yourself out of your comfort zone so much and you experience so much. You may find yourself creating a different path, a new attitude, or a new mindset from this experience. Look at me. I went from being a forecourt station manager in Ireland to becoming an English teacher and working in the best school in the province of Guizhou, becoming a certified PE teacher and an international spinning instructor in China. I do not mind telling you that there are many people I know, even some family members and friends who said and thought that I would never amount to anything, and that they could never see me hacking China and that the culture shock would be too much to handle, and I would return home. Well, I am so happy to say six years later, I could not be happier, it is the reverse culture shock that I have problems with when I go home.

Lastly, if you ever find yourself in a rut or a situation (especially related to work), remember only you have the power to change your situation. Nobody else. I did it. Just re-read the chapters I spent discussing my career change and seek the inspiration from that. I will end my story now with those questions and statements that I asked myself as I sat on the cliffs looking down upon The Treasury in Petra, those questions that resulted in me travelling all of Asia and living the life I thought I could never have.

Remember this, you can do it, you need to want to do it. "What are you doing with your life?"

"You need to do more of this, because this is what you love deep inside."

"It is time to say fuck off to the 9–5 job you have, you need more from life."

"You need more goals than just hitting KPIs for a job and manager that would replace you in the morning for not hitting them."

"Why settle for just a piece of sky?"

Finally, I want to tell you why I decided to call this book 'Squat Toilets and Chopsticks'. You probably think it is a peculiar name for a book.

I would like state that under no circumstances is the choice of wording for the title meant to cause any offense to anyone of any culture.

The title mentions two things which I learned to use in China. One that I can use like a local person and one thing that after three years I still cannot use! And yes, it is the chopsticks that I have perfected in using! The squat toilet, not so much!

But there is another meaning, a much deeper meaning behind the title. I decided to pay homage to my favourite movie, Disney's *Bedknobs and Broomsticks*. I have watched this movie since I was two years old, and it is special to me. I still watch it at least once a month. The movie is set in England during World War II, and it tells the story of an apprentice witch's efforts to assist against the enemy. It takes her, a professor and three children off on a magical adventure on a flying bed. Together, they travel to magical places in search of a spell that could help end the war. A spell that can make objects come to life.

This special movie has taught me to believe in magic and that anything is possible. It was also the very first movie I saw that featured travelling, except in the movie it is a flying bed and not an airplane. Now a flying bed would be fantastic if it existed today! The movie has served as inspiration to me over the years and as comfort for when I feel down. It reminds me of that magic.

www.ingramcontent.com/pod-product-compliance
Lightning Source LLC
Chambersburg PA
CBHW070239190526

45169CB00001B/230

9 7 8 1 5 2 8 9 5 3 3 2 0